Higher BIOLOGY

For SQA 2019 and beyond

Revision + Practice
2 Books in 1

001/01012020

10 9 8 7 6 5 4 3 2 1

ISBN 9780008365288

Published by
Leckie & Leckie Ltd
An imprint of HarperCollins*Publishers*
Westerhill Road, Bishopbriggs, Glasgow, G64 2QT
T: 0844 576 8126 F: 0844 576 8131
leckieandleckie@harpercollins.co.uk www.leckieandleckie.co.uk

Publisher: Sarah Mitchell
Project Manager: Lauren Murray

Special thanks to
Jouve (layout and illustration)
Louise Robb (proofreading)

Printed and bound by CPI Group (UK) Ltd, Croydon CR0 4YY

A CIP Catalogue record for this book is available from the British Library.

Acknowledgements

P36 Enzyme structure © J.C. REVY, ISM/SCIENCE PHOTO LIBRARY; P30 Intestine archaea Methanobrevibacter smithii © SEM CREDIT: DENNIS KUNKEL MICROSCOPY/SCIENCE PHOTO LIBRARY; P73 Illustration of a tapeworm © SEBASTIAN KAULITZKI/SCIENCE PHOTO LIBRARY

All other images © Shutterstock.com

'*For all my pupils in Orkney, past and present.*' Angela Drummond

ebook

To access the ebook version of this Success Guide visit
www.collins.co.uk/ebooks
and follow the step-by-step instructions.

ANSWERS Check your answers to the practice test papers online:
www.leckieandleckie.co.uk

Higher Complete Revision and Practice

Complete Revision and Practice

This Complete two-in-one Revision and Practice book is designed to support you as students of Higher Biology. It can be used either in the classroom, for regular study and homework, or for exam revision. By combining a revision guide and two full sets of practice papers, this book includes everything you need to be fully familiar with the Higher Biology exam. As well as including ALL the core course content with practice opportunities, there is assignment and exam preparation advice; links to the syllabus to help with targeted topic practice and easy reference with a glossary and Quick Test answers.

How to use the revision guide

This revision guide is designed to help you in both your revision work for Higher Biology, and to support you with current learning. The topics are broken down into concise areas that will assist you with recall and help you to lock in key points. Top Tips in each chapter guide you towards success and help you to remember key facts. Quick Tests at the end of each chapter are designed to ensure that you have learned essential facts in order to improve your performance in exam questions that test knowledge and understanding. Use the Quick Tests to regularly check that you've retained the important information on each topic.

Answers to Quick Tests can be found at the back of the revision guide.

Throughout the book, examples are given, where possible, to illustrate biological concepts. Use of examples is an important element for success in extended response questions.

The guide follows the area structure of the course as detailed in the SQA National Course Specification Document. It's advisable to use the revision guide alongside other resources: it is not a one-stop route to exam success but should be an important tool in building up your knowledge and understanding of the course and your confidence in your abilities to apply that learning.

Success in Higher Biology also depends on your knowledge of new vocabulary. Many assessment questions test your knowledge of the meaning of a particular word or phrase. To help you learn words and meanings, a glossary of terms has been included at the end of the revision guide.

Use the revision guide as a starting point – come back to it for the essential knowledge and skills that you need to tackle questions and understand concepts and processes in biology. This revision guide may also provide useful support during biology lessons, and can be used as a quick reference book in addition to other resources.

How to use the practice exam papers

This book contains two practice papers, which mirror the actual SQA exam as closely as possible in question style, level and layout. It is the perfect way to familiarise yourself with what the exam papers you will sit will look like.

The answer section, which you can find online at www.leckieandleckie.co.uk, contains worked answers to the questions which appear in the practice papers, letting you know exactly where the marks are gained in an answer and how the right answer is arrived at. It also includes practical tips on how to tackle certain types of questions, details of how marks are awarded and advice on just what the examiners will be looking for.

The practice papers can be used in one of two ways:

1. You can complete an entire practice paper under exam conditions and mark it using the answer section. If you complete a practice paper this way it is important to make a list of all the key areas that you are having difficulty with by referring to the links to the syllabus on pages 100–101, and concentrating your study time on these areas before attempting the next practice paper.
2. You can use the links to the syllabus to target specific topics you may wish to revise, by selecting only those questions on those specific topics.

The Higher Biology course

The Higher Biology course offers relevant, up-to-date information on a broad range of topics in biological science, which provide the foundation for biological research today.

The course covers key principles of biology including molecular biology, biochemistry, physiology, genetics, plant science, zoology, biodiversity and agriculture. There is a great deal of exciting biology to learn within the course, and regular weekly revision of work is advisable in order to strengthen knowledge and understanding.

Structure

The Higher Biology course is made up of three areas:

- DNA and the genome
- Metabolism and survival
- Sustainability and interdependence

Course assessment

The course assessment takes the form of two question papers that form the National Examination and a course assignment.

The exam

The National Examination gives you the opportunity to apply skills in both problem-solving and knowledge and understanding, answering questions from across the three areas of the course.

The exam carries a total of 120 marks, represents 80% of the total marks for the course and is made up of two sections:

- Paper 1 (25 marks) comprising 25 multiple-choice questions lasting 40 minutes
- Paper 2 (95 marks) made up of a mixture of restricted and extended response questions; there will be two extended response questions (totalling around 9–15 marks), one large question based on an experiment (worth around 5–9 marks) and one large data-handling question (worth around 5–9 marks) lasting 2 hours 20 minutes

You should be aware that techniques such as gel electrophoresis and thin layer chromatography as well as apparatus, including respirometers, can be assessed here as well.

The assignment

The assignment assesses skills such as handling and processing data gathered through experimental work and research skills. It is worth 20 marks and represents 20% of the overall marks for the course assessment. It is recommended that no more than 8 hours are spent on the assignment with a maximum of 2 hours on the report stage.

There are two stages to the assignment:

1. Research stage:

Under some degree of supervision and control you will:

i) plan, design and carry out an experiment in class that is relevant to your chosen topics aim and collect data.

ii) research the underlying biology of your chosen topic and find another source of data relevant to your experiment from an internet/literature source.

2. Report stage:

Here you will write a structured report of your findings under examination conditions. Your report will include relevant underlying biology, experimental data, analysis and comparison of your data with data from your other internet/literature source. Your report will contain an evaluation and allow a valid conclusion that relates to the aim to be drawn.

Higher BIOLOGY

For SQA 2019 and beyond

Revision Guide

John Di Mambro, Angela Drummond and Deirdre McCarthy

The structure of DNA

Genetic information within every living cell is contained within deoxyribonucleic acid or DNA. It is a ladder-shaped molecule, twisted to form a **double helix**. A DNA molecule is composed of two strands, each of which is made up of repeating units called nucleotides.

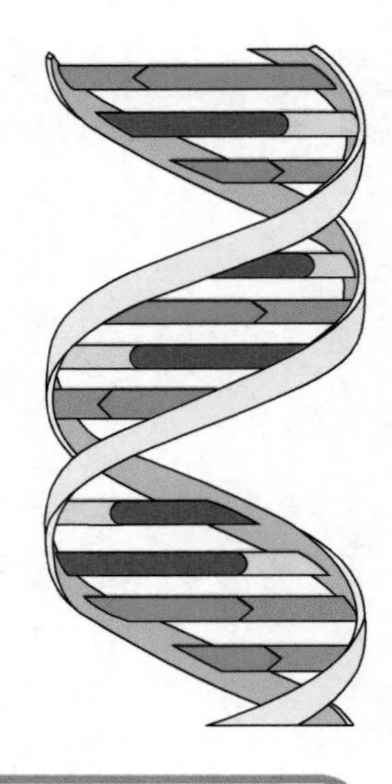

Nucleotide structure

A nucleotide consists of a phosphate group, deoxyribose sugar and a base. The deoxyribose sugar molecule has five carbon atoms. Carbon atom 1 is on the right, then count anti-clockwise to carbon atoms 2, 3 and 4, with carbon atom 5 situated between carbon atom 4 position and the phosphate group.

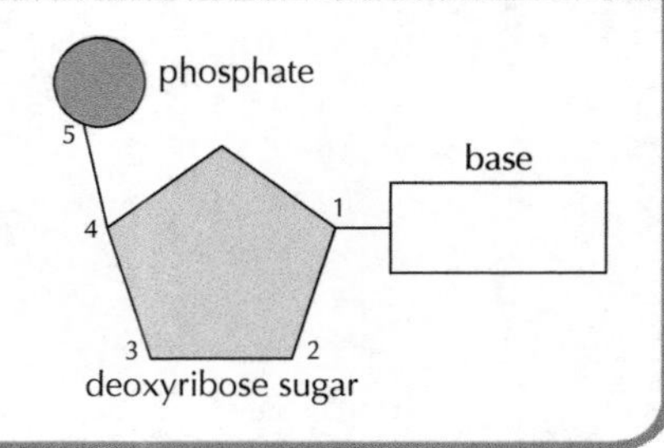

Double stranded and antiparallel

TOP TIP

Remember: the DNA backbone consists of two components 1) phosphate group and 2) deoxyribose sugar.

The phosphate group of one DNA nucleotide and deoxyribose sugar of the next are joined by a strong bond. Both of these components make up the sugar-phosphate backbone of each strand of the DNA molecule.

DNA is double-stranded. The two strands run in opposite directions from the deoxyribose (3′) at one end to phosphate (5′) at the other end. Thus, we describe DNA as being double stranded and **antiparallel**.

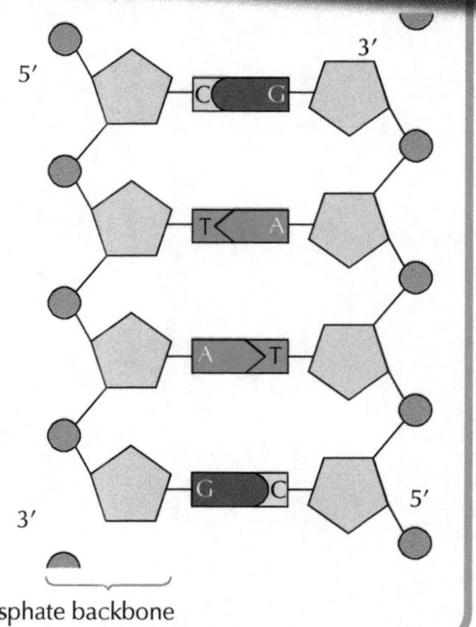

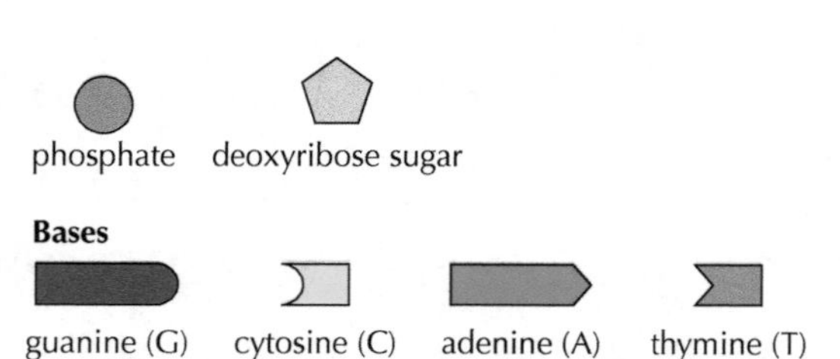

DNA base pairing

DNA has four bases that combine in pairs to form the 'rungs' of the ladder-shaped molecule. Pairing always follows the same rule: adenine pairs with thymine, and cytosine pairs with guanine.

Complementary base pairs are held together by hydrogen bonds.

The nucleotide base sequence of the DNA molecule forms the genetic code.

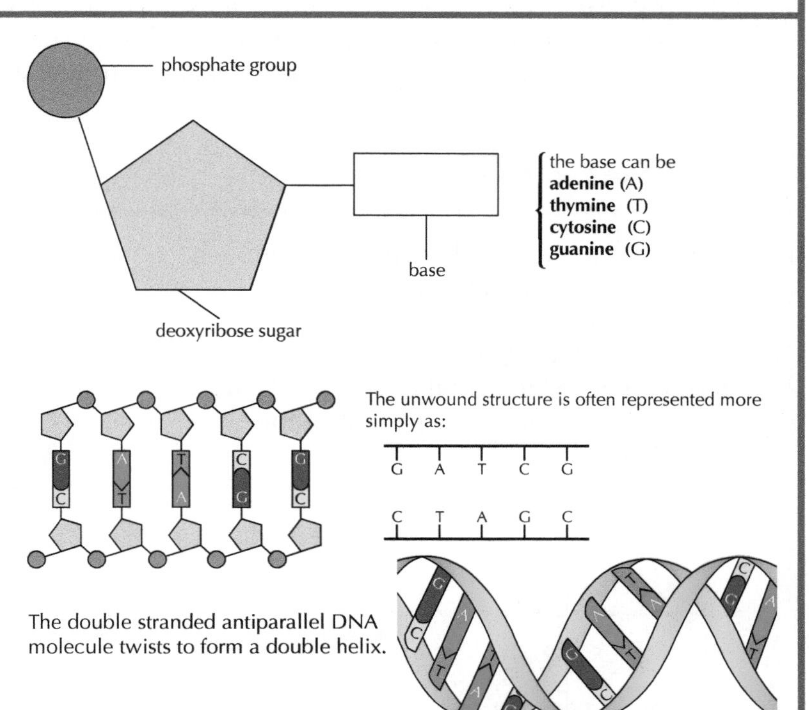

TOP TIP

Remember: base pairing rule:

- straight-edged letters go together, A and T
- curved letters go together, G and C.

Remember, you only need to know the percentage of one base in a strand to be able to calculate the percentages of the other three.

Quick Test 1

1. Name the two components of the DNA backbone.
2. Explain what is meant by the term antiparallel.
3. Describe what is meant by the term genetic code.

Organisation of DNA

In both **prokaryotic** and **eukaryotic** cells, DNA is organised into structures called **chromosomes**. In prokaryotes (such as bacteria) chromosomes are circular. In eukaryotes (such as animal and plant cells and yeast) chromosomes are linear.

Prokaryotic cells

A prokaryotic cell, such as a bacterium, does not have a membrane-bound nucleus. DNA is found in the cytoplasm in the form of a large, single circular chromosome. Many smaller rings of DNA called **plasmids** are also present in the cytoplasm.

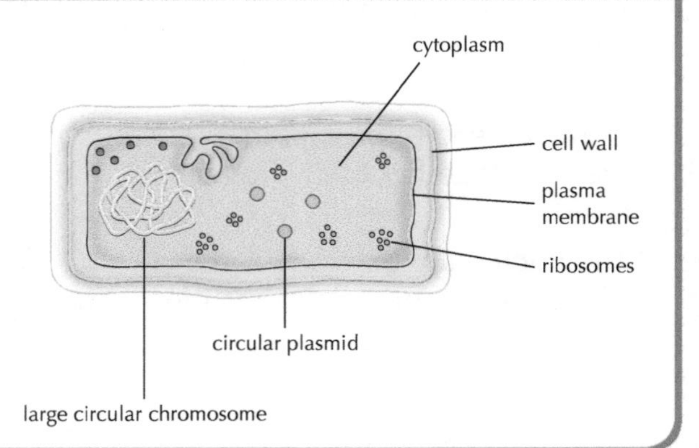

Eukaryotic cells (plant and animal)

A eukaryotic cell has a membrane-bound nucleus containing linear chromosomes. The DNA molecule that makes up a chromosome can be several metres long, and so it is tightly coiled and packaged, around associated proteins called **histones**.

Small, circular chromosomes are found in the **chloroplasts** and **mitochondria** of plant cells, and the mitochondria of animal cells.

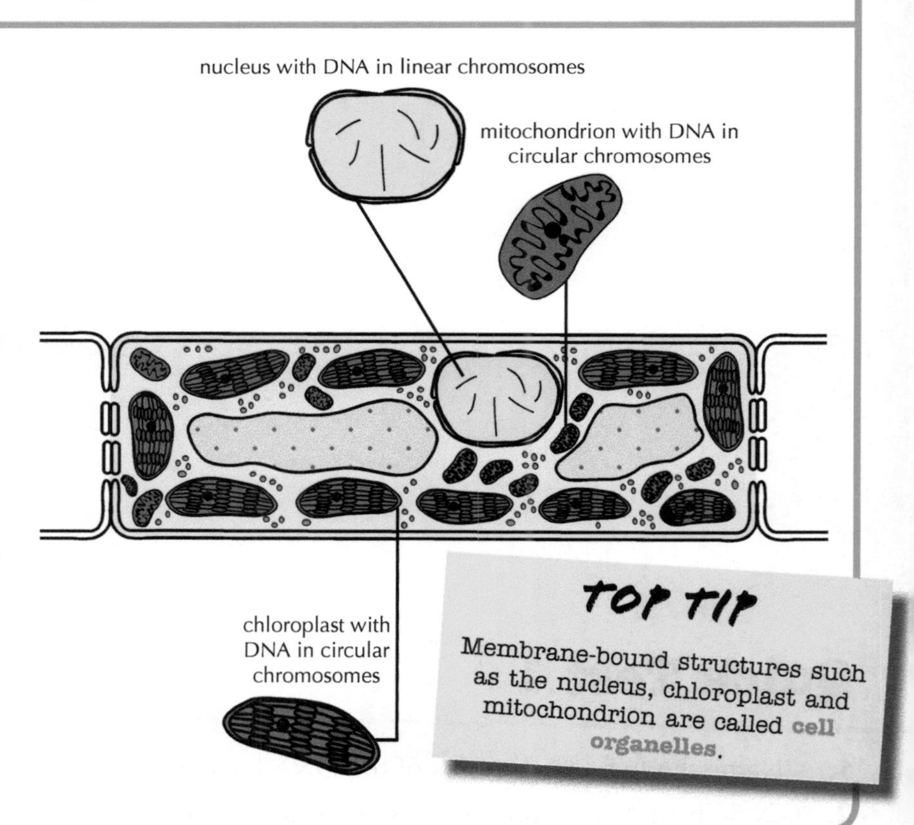

TOP TIP

Membrane-bound structures such as the nucleus, chloroplast and mitochondrion are called **cell organelles**.

GOT IT? ☐ ☐ ☐

Chromosomes condense in eukaryotic cells during cell division, when they replicate and become visible with a microscope.

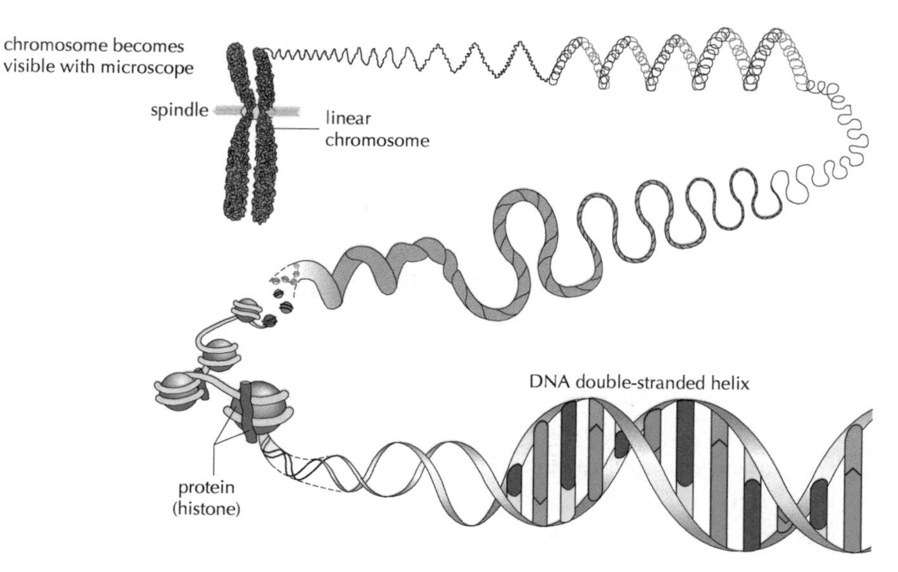

Eukaryotic cells (yeast)

Yeast cells are eukaryotic and are a type of single-celled fungus. They contain linear chromosomes within a membrane-bound nucleus. They are unusual as they also contain some circular DNA plasmids within their cytoplasm, similar to those found in prokaryotes.

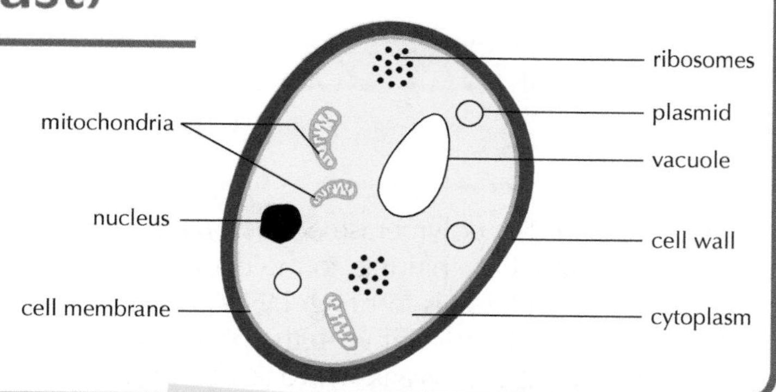

TOP TIP

Bacterial and yeast plasmids are used in genetic engineering. They are extracted from bacteria and yeast cells, cut open with **restriction endonuclease** enzymes leaving 'sticky ends'. A gene from another organism is placed between the '**sticky ends**' and the plasmid is sealed with **ligase** enzyme. The recombinant bacterial and yeast cells now have genes that can code for foreign proteins.

TOP TIP

Cell measurements:
1 mm = 1000 micrometres (μ)
Multiply mm × 1000 to get micrometres.
Remember: watch out for questions that look for the actual size of a cell. The magnification tells you how many times bigger the cell looks under the microscope. For example, a magnification of ×400 makes what you see under the microscope appear 400 times bigger than it really is. Divide your calculated cell size by the magnification to get the actual size of the cell.

Quick Test 2

1. State how the organisation of DNA is different in a prokaryote compared with a eukaryote.
2. Name the small circular DNA structure found in both prokaryotes and yeast cells.
3. State the form taken by DNA in a chloroplast.

Replication of DNA

Copying the genetic code

The DNA in the nucleus of a cell must make an exact copy of itself (replicate) before a cell can divide by mitosis. The DNA molecule acts as a **template** for DNA replication.

Initially, the DNA double helix unwinds and the hydrogen bonds between bases break. The two strands unwind steadily to reveal the two template strands. DNA is replicated by an enzyme called **DNA polymerase**.

A primer is a short strand of nucleotides that binds to the 3′ end of the DNA template strand. A primer is needed in order for DNA polymerase to start replication.

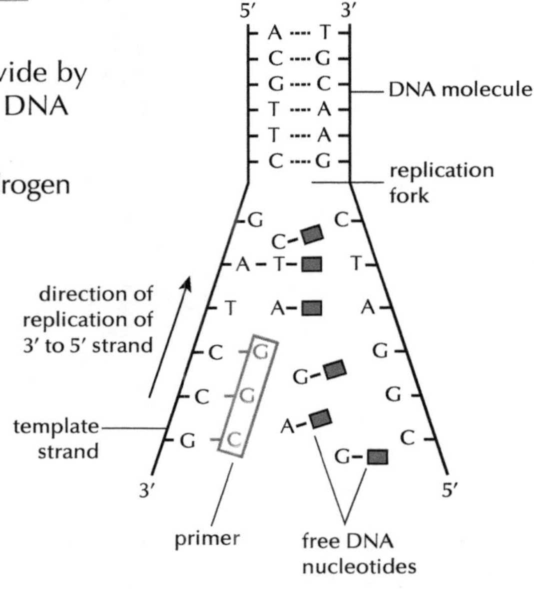

The stages of DNA replication

TOP TIP

DNA polymerase adds DNA nucleotides, using complementary base pairing, to the deoxyribose (3′) end of the primer and thereafter to the deoxyribose (3′) end of the newly forming strand.

The enzyme DNA polymerase adds DNA nucleotides, using complementary base pairing, to the deoxyribose 3′ end of the new DNA strand that is forming. DNA polymerase also forms strong chemical bonds between the phosphates and sugars of the new DNA strand. Weak hydrogen bonds form between the base pairs of the new strand and the template strand.

DNA polymerase can only add nucleotides in one direction. As a result, the **leading strand** is replicated continuously.

Primers are added to the lagging strand as it is unwound at the **replication fork**. Thus, the **lagging strand** is replicated in fragments.

Fragments of DNA are then joined together by the enzyme ligase.

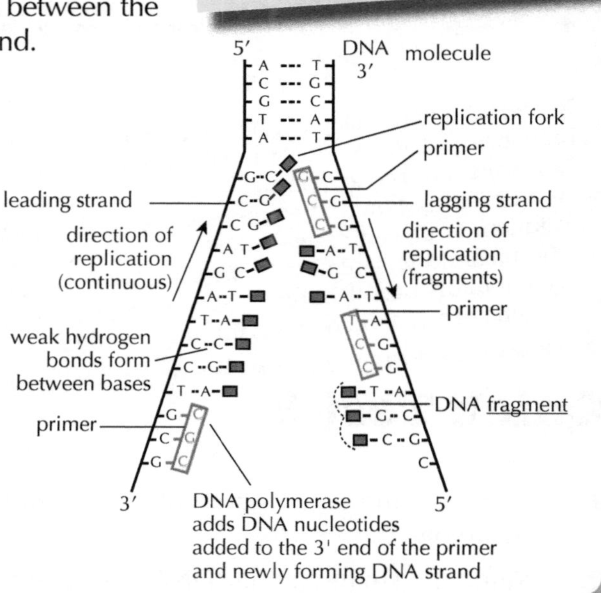

Leading strand

Replication of DNA from the 3′ end is continuous, moving towards the junction of the replication fork; this is called the leading strand.

Lagging strand

Replication from the 5′ end is discontinuous, moving away from the replication fork; this is called the lagging strand. The lagging strand is replicated in fragments.

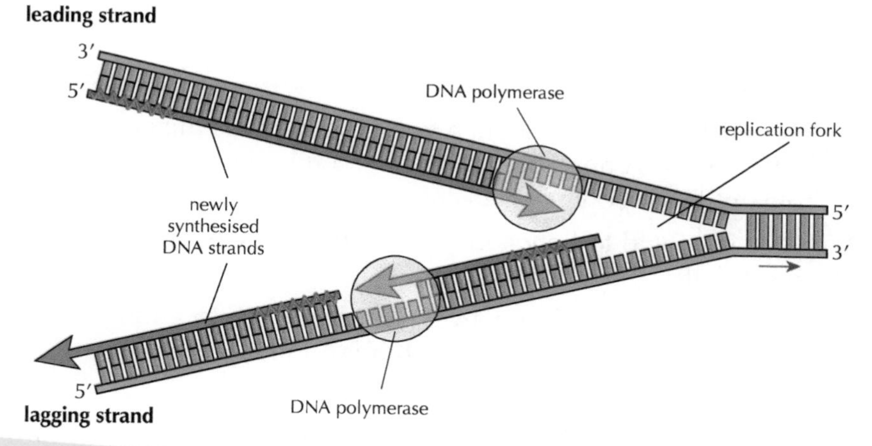

TOP TIP

Remember: for DNA to replicate, a nucleus must contain:
1. DNA to act as a template
2. four types of free DNA nucleotides
3. ATP
4. primers
5. enzymes – DNA polymerase and ligase.

DNA polymerase and ligase

1. DNA polymerase is an enzyme that adds DNA nucleotides to the 3′ end of a primer in one direction only.
2. Primers are needed so that DNA polymerase can start replication.
3. Ligase joins fragments of DNA together.

Quick Test 3

1. State the function of a primer.
2. State the end of the primer to which DNA nucleotides are added.
3. Name the enzyme that joins DNA fragments together.
4. Explain why the lagging strand is replicated in fragments.

Polymerase chain reaction (PCR)

Tiny fragments of DNA may be copied to provide enough for analysis. This is called **amplification** of DNA, and is done using a process called the **polymerase chain reaction (PCR)**. One cycle of PCR involves three steps carried out at three different temperatures, a process called **thermocycling**. In PCR, primers are short strands of nucleotides that are complementary to specific target sequences at the two ends of the region of DNA to be amplified.

The stages of PCR

1. The specific target sequence to be amplified is mixed with free DNA nucleotides in a small tube and heat tolerant DNA polymerase is added.
2. DNA is heated to between 92 and 98°C to separate the strands.
3. DNA is then cooled to between 50 and 65°C to allow the complementary primers to bind to the target sequences.
4. DNA is then heated to between 70 and 80°C for the heat tolerant DNA polymerase to replicate the target sequence.
5. Repeated cycles of heating and cooling amplifies the target sequence.

The DNA fragment to be copied goes through 30 cycles per hour, making one million copies of the original fragment for analysis!

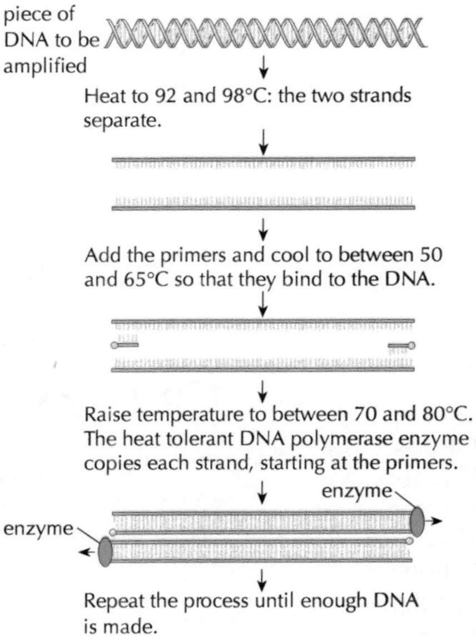

piece of DNA to be amplified

Heat to 92 and 98°C: the two strands separate.

Add the primers and cool to between 50 and 65°C so that they bind to the DNA.

Raise temperature to between 70 and 80°C. The heat tolerant DNA polymerase enzyme copies each strand, starting at the primers.

enzyme

enzyme

Repeat the process until enough DNA is made.

Top tip

Examination questions are likely to ask about the temperatures used in PCR, and what happens at each temperature:

- 92–98°C – to separate DNA strands
- 50–65°C to allow complementary primers to bind to target sequences
- 70–80°C for heat tolerant DNA polymerase to replicate the region of DNA.

HEAT–COOL–HEAT–REPEAT

Top tip

Remember: The target sequence to be amplified is double stranded so two specific complementary primers are needed to target each strand.

Primers are short strands of DNA nucleotides that allow polymerase to bind to their 3′ end.

Practical applications of PCR

PCR produces millions of copies of (amplifies) the target sequence. These copies provide enough DNA for further analysis. The amplified sample of DNA can then be used to:

1. help solve crimes
2. help solve paternity suits
3. diagnose genetic disorders.

Gel electrophoresis separates molecules

When copies of a DNA fragment have been made, some may be treated with restriction endonuclease enzymes that cut the fragments into smaller pieces at specific base sequences. These fragments are separated out by gel electrophoresis to form a pattern of bands. This pattern is called a DNA fingerprint and is used to compare samples of DNA, which may or may not have come from the same person.

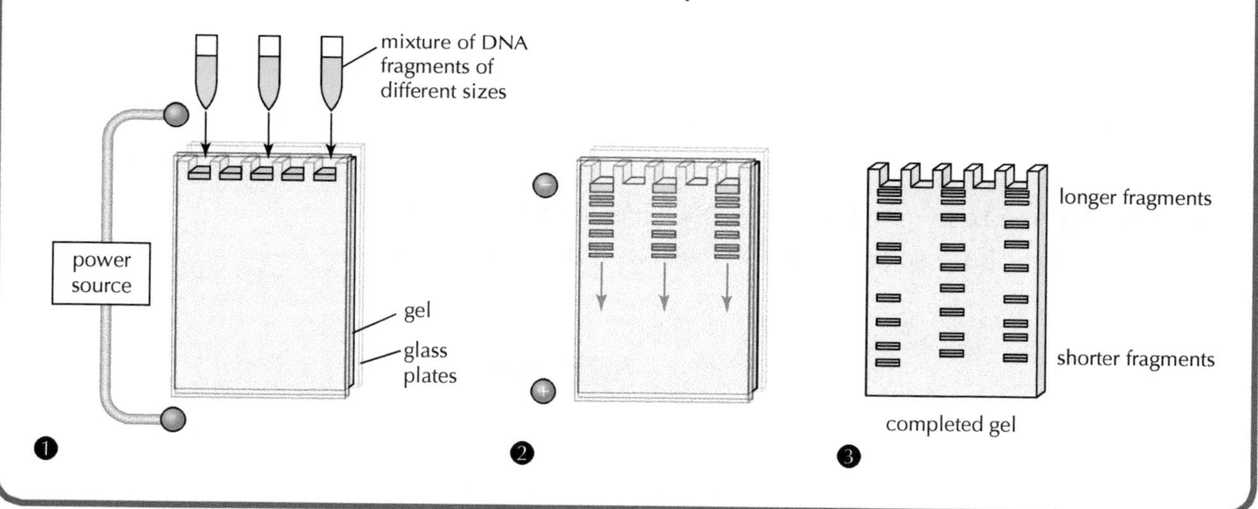

Quick Test 4

1. Explain why heat tolerant DNA polymerase is used in PCR.
2. State two practical applications for the DNA amplified by PCR.
3. State one use of gel electrophoresis.

Gene expression

Gene expression and phenotype

The **phenotype** of an organism is determined by proteins produced as a result of **gene expression**.

Gene expression involves the processes of **transcription** and **translation**. These processes turn the genetic code found in the DNA base sequence of a gene into a sequence of amino acids and ultimately a protein.

Only a fraction of the genes in a cell are expressed at any one time.

Eukaryotic cells continually switch genes 'on' and 'off' in response to signals from inside and outside the cell. If a gene is switched on it is coding for a protein. Only some genes within a cell are switched on or 'expressed' at any one time.

TOP TIP

Gene expression has 3 main stages:

- transcription
- **RNA splicing**
- translation.

Ribonucleic acid (RNA)

This is a second nucleic acid involved in gene expression, which differs from DNA as follows:

- RNA has the base uracil whereas DNA has the base thymine.
- RNA has a ribose sugar whereas DNA has a deoxyribose sugar.

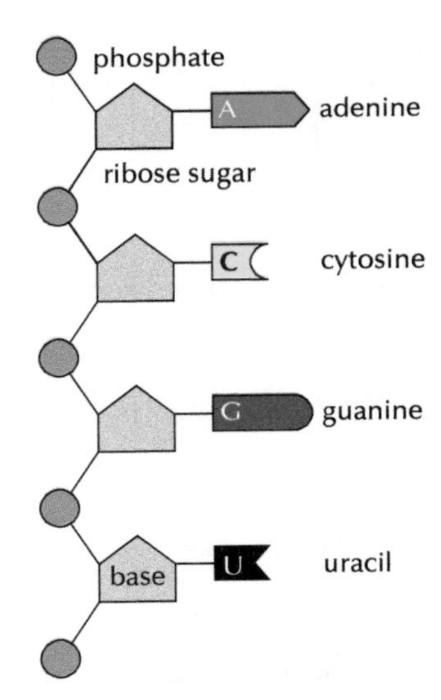

Types of RNA

Gene expression involves three types of RNA (mRNA, tRNA and rRNA):

1. Messenger RNA (mRNA) carries a copy of the DNA code from the nucleus to the **ribosome**. Each triplet of bases on the mRNA molecule is called a codon and codes for a specific amino acid. Messenger RNA is transcribed from DNA in the nucleus and is translated into proteins by ribosomes in the cytoplasm.
2. Transfer RNA (tRNA) folds due to complementary base pairing. Each tRNA molecule carries a specific amino acid to the ribosome. A tRNA molecule has an anticodon (an exposed triplet of bases) at one end and an attachment site for a specific amino acid at the other end.
3. Ribosomal RNA (rRNA), along with ribosomal proteins, forms a ribosome.

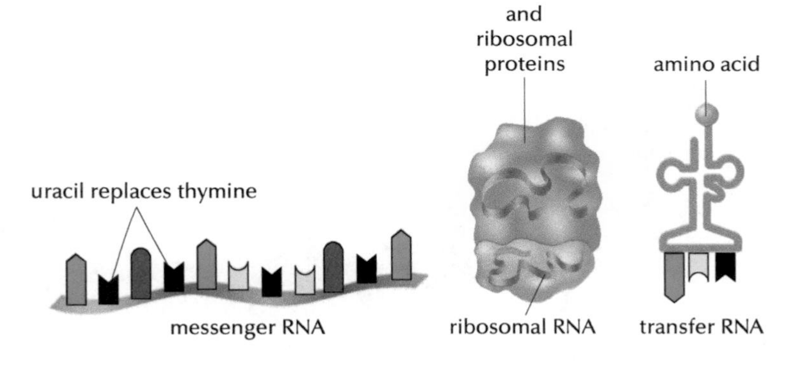

Introns and exons

When mRNA is transcribed in the nucleus, a primary mRNA transcript is formed.

Introns and **exons** are found in the primary mRNA transcript.

1. The introns of the primary transcript are non-coding regions.
2. The exons of the primary transcript are coding regions.

TOP TIP

Remember: Exons are expressed! They actively code for a protein.

Quick Test 5

1. Name two structural differences between DNA and RNA molecules.
2. State what determines the phenotype of an organism.
3. Name the three components of an RNA nucleotide.
4. State what makes up a ribosome.
5. State the difference between exons and introns.

The process of copying the base sequence (DNA code) of a gene occurs in the nucleus and is called transcription. DNA in the nucleus acts as a template for the production of a **primary transcript** of mRNA. DNA cannot leave the nucleus of a cell. It is the mature mRNA transcript that carries a complementary copy of the DNA from the nucleus to the ribosome in the cytoplasm.

Gene expression – transcription

1. The enzyme, **RNA polymerase**, moves along the DNA unwinding the double helix and breaking the hydrogen bonds between bases.
2. RNA polymerase synthesises a primary transcript of mRNA from RNA nucleotides by complementary base pairing.

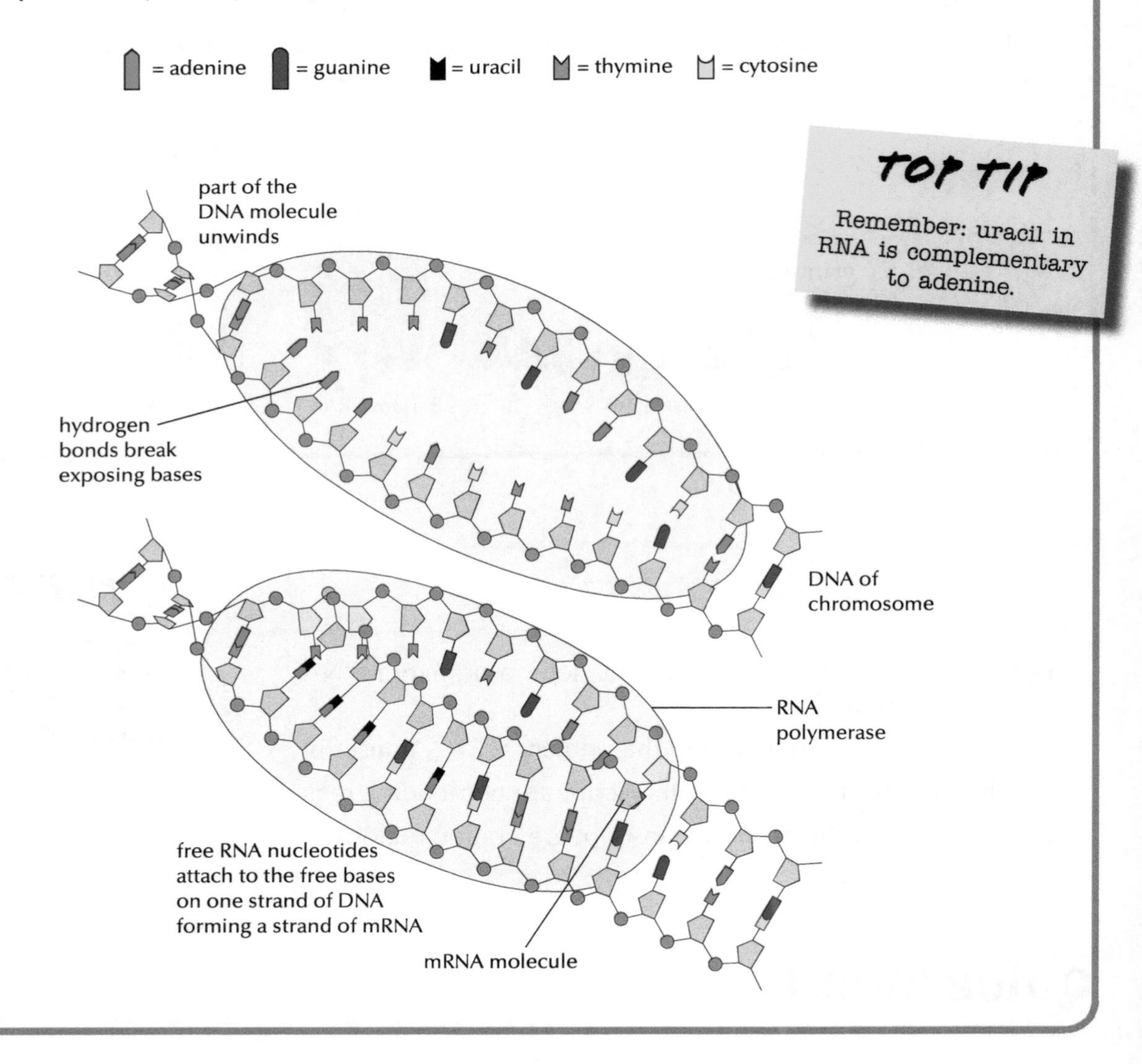

Gene expression – mRNA splicing

RNA splicing forms a mature mRNA transcript.

1. The primary mRNA transcript is composed of introns and exons. The introns of the primary transcript are non-coding regions and are removed.
2. The exons are coding regions and are joined together to form the mature transcript.
3. The order of the exons remains unchanged during splicing.
4. The mature mRNA transcript, formed as a result of RNA splicing, leaves the nucleus and travels to a ribosome in the cytoplasm for the next stage in gene expression.

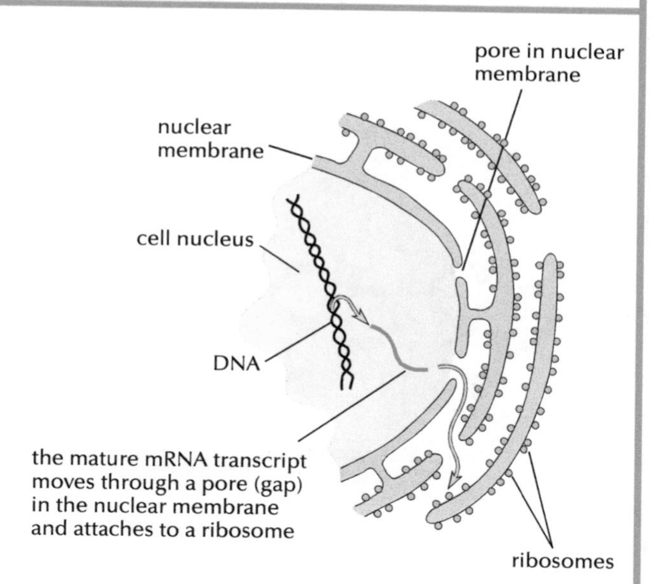

TOP TIP

Be prepared for questions that involve transcribing a DNA sequence into an mRNA sequence. Be able to explain why there is a difference in length between primary and mature mRNA transcripts. Remember: the primary transcript contains exons and introns but the mature transcript contains only exons.

TOP TIP

Ribosomes can be found free in the cytoplasm or attached to membrane surfaces.

Quick Test 6

1. If part of the DNA base sequence on a gene is:
   AAG CGT GGT ATG ACC
   State the complementary base sequence of the mRNA that would be transcribed from this DNA base sequence.
2. State the function of RNA polymerase.
3. Describe the formation of a mature mRNA transcript.
4. Name the organelle to which the mRNA becomes attached after it leaves the nucleus and enters the cytoplasm.

Gene expression – translation

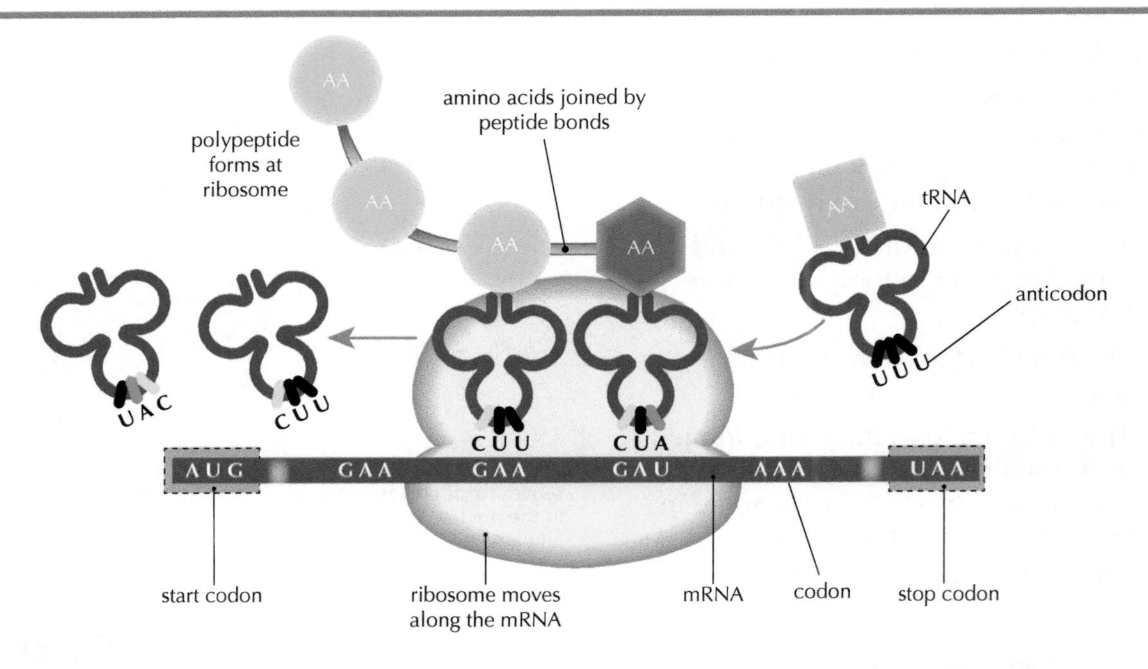

Translation is the part of gene expression where the mature mRNA transcript is used to synthesise a **polypeptide** chain at the ribosome.

There are a number of stages in this translation process.

1. mRNA attaches to the ribosome.
2. Translation begins at a start codon and ends at a stop codon.
3. tRNA has an anticodon at one end and an attachment site for a specific amino acid at the other end.
4. The mRNA codons bond to tRNA anticodons by complementary base pairing.
5. The genetic code is translated into a sequence of amino acids.
6. Peptide bonds join the amino acids together.
7. Each tRNA then leaves the ribosome as the polypeptide is formed.
8. The last codon on the mRNA is called a stop codon and this ends the translation process.

TOP TIP

Remember: gene – DNA base sequence – mRNA codons – tRNA anticodons – amino acid sequence.

TOP TIP

Polypeptide chains can be arranged to form different proteins such as enzymes, antibodies, hormones and structural proteins.

Structure of proteins

Protein shape

Proteins have a large variety of shapes that determine their function.

The polypeptide chain produced as a result of gene expression undergoes a process of folding to form the three-dimensional shape of a protein.

Folding occurs due to hydrogen bonds and other interactions between individual amino acids that link polypeptide chains together.

The sequence of amino acids and the interactions between polypeptide chains determines the 3D shape of the protein, which in turn determines its function.

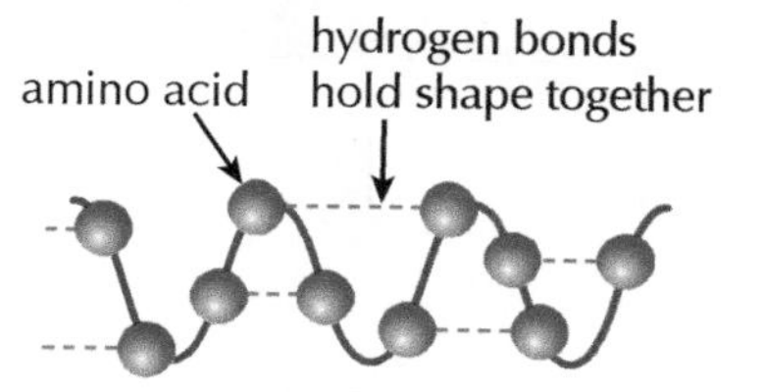

Quick Test 7

1. Describe the structure of an anticodon.
2. State how many amino acids can be carried by a tRNA molecule.
3. State the type of chemical bond that joins amino acids together to form a polypeptide chain.
4. Explain the function of a 'stop codon'.

GOT IT? ☐ ☐ ☐

Alternative RNA splicing

Different mature mRNA transcripts can be produced from the same primary transcript depending on which exons are retained. In this way, one gene can code for more than one protein.

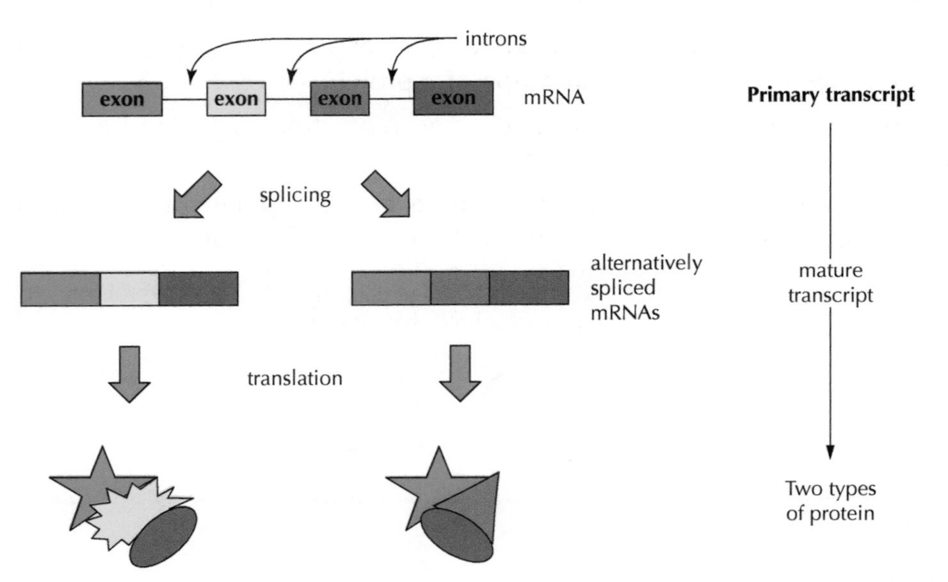

Quick Test 8

1. State the term used to describe the coding regions of a primary mRNA transcript.
2. A double-stranded fragment of DNA that contains the gene for lactase contains no introns and 2976 nucleotides. Calculate the number of amino acids in the enzyme.
3. Name and describe the process that results in different mRNA molecules being expressed.

Differentiation in cells

Cellular **differentiation** is the process by which a cell expresses certain genes to produce proteins characteristic of that cell type. This allows a cell to carry out specialised functions.

Meristems in plants

Meristems are regions of unspecialised cells in plants that can:

1. divide (**self renew**) and/or
2. differentiate.

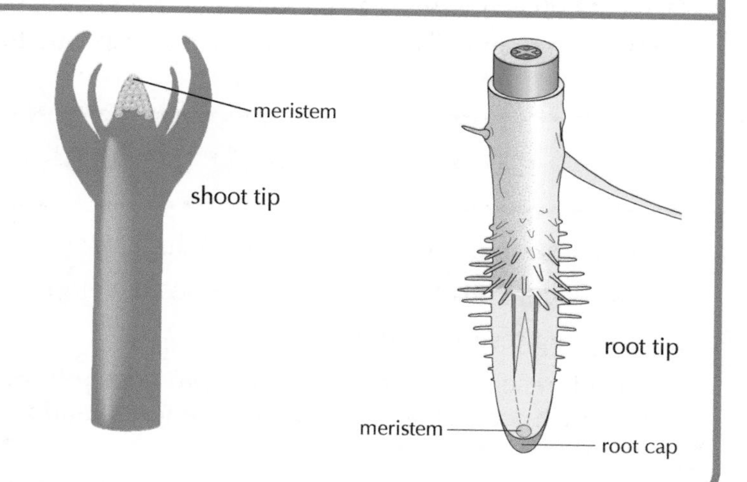

TOP TIP

Remember from National 5: specialisation of cells leads to the formation of a variety of cells, tissues and organs. The cells in multicellular organisms are specialised for their function and work together to form systems.

Stem cells in animals

Stem cells are unspecialised cells in animals that can divide (self-renew) and/or differentiate.

There are two types of animal stem cells:

1. Tissue stem cells.
2. Embryonic stem cells.

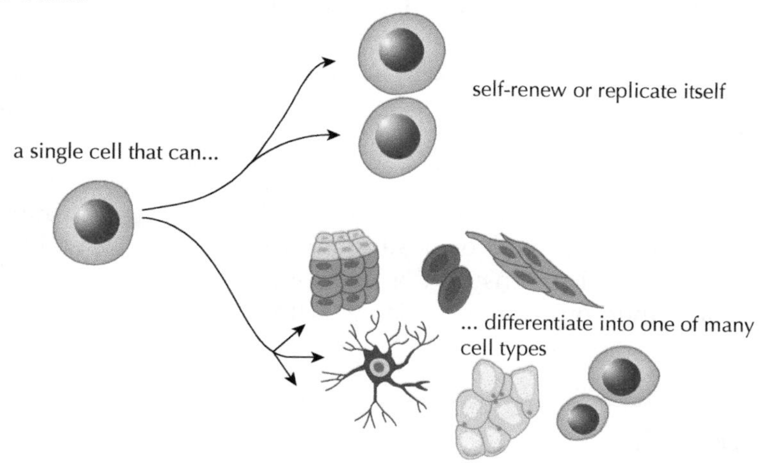

Tissue stem cells

These are involved in the growth, repair and renewal of cells found in a particular tissue. They are **multipotent**. This means that they can differentiate into all of the cell types found in a particular tissue. For example: blood stem cells located in bone marrow can give rise to all types of blood cell.

GOT IT?

Embryonic stem cells

Stem cells isolated from a human embryo can self-renew under the right conditions in the laboratory.

Stem cells found in the very early embryo can differentiate into all the cell types that make up the organism and so are called **pluripotent**. All the genes in embryonic stem cells can be switched on so these cells can differentiate into any type of cell.

Embryonic stem cells can offer effective therapeutic treatment for both disease and injuries.

Stem cells – research uses

Current research uses of stem cells include use:

1. as model cells to study how diseases develop
2. for drug testing in the laboratory.

Stem cell research is important as it provides information on how cell processes such as cell growth, cell differentiation and gene regulation work.

Stem cells – therapeutic uses

Uses of stem cells in medicine and therapeutics includes repair of diseased organs or tissues.

Tissue stem cells are currently used in corneal repair to replace damaged corneal tissue and thus restore sight.

Tissue stem cells are also used to regenerate damaged skin following a burn. New skin can be grown in the laboratory from the stem cells of a patient with a burn and grafted back onto the damaged area without fear of rejection.

Ethical issues

The use of embryonic stem cells raises ethical issues. These include the fact that it involves destruction of embryos. Currently regulations ensure that the use of stem cells in research and therapy is in accordance with UK law.

TOP TIP

Don't forget!
Multipotent = can differentiate into all the types of cell found in a particular tissue type.
Pluripotent = can differentiate into all the cell types that make up the organism.

Quick Test 9

1. Define cellular differentiation.
2. State one fact about the potential for gene expression in embryonic stem cells.
3. Name a source of tissue stem cells in humans.
4. State one research use of human stem cells.

Structure of the genome

The genome

The genome of an organism is its entire hereditary information encoded in DNA.

The genome is made up of genes and other DNA sequences that do not code for proteins.

Function of 'non-protein coding sequences' or introns

Genes are defined as DNA sequences that code for protein; however, most of the eukaryotic genome consists of non-coding sequences.

Some of these non-coding sequences regulate transcription and others are transcribed into another molecule but never translated into protein. tRNA and rRNA are examples of very useful non-translated forms of RNA.

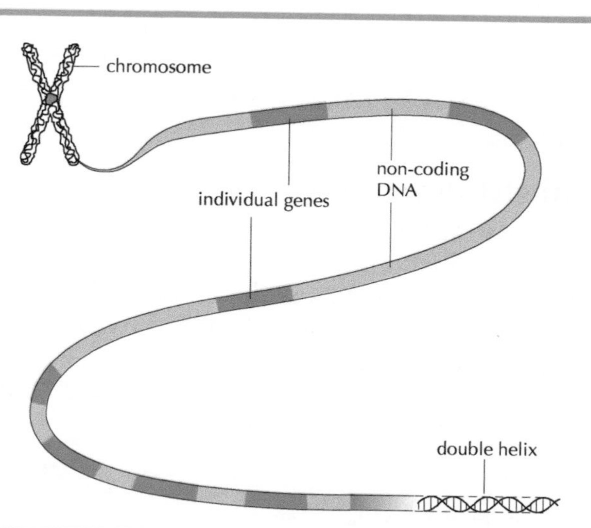

Mutation

A mutation is an irreversible change in the sequence of nucleotides within a gene or chromosome, resulting in either no protein or an altered protein being synthesised.

A change in **genotype** due to a mutation causes a change in phenotype of the organism, which is referred to as a **mutant**.

Mutation provides a source of genetic variation on which the process of evolution depends. Mutations cause alterations to the genome that may contribute to the evolution of the **species**.

Mutations can occur during DNA replication or gamete formation.

There are two main categories of mutations:

1. single gene mutations.
2. chromosome structure mutations.

TOP TIP

Make sure you know the definition: a mutation is a change in the DNA that can result in no protein or an altered protein being synthesised.

TOP TIP

Remember mutations occur randomly and at low frequency.

Single gene mutation

The alteration of a DNA nucleotide sequence as a result of the **substitution**, **insertion** or **deletion** of nucleotides is known as a single gene mutation.

Single gene mutations can be categorised as:

a) substitution mutations:

i. **missense**

ii. **nonsense**

iii. **splice-site**

b) **insertion** – (**frame-shift** effect)

c) **deletion** – (**frame-shift** effect)

Substitution mutations

TOP TIP

You do not need to know which codons code for each amino acid.

Substitution – missense

One base within a triplet of DNA bases is substituted for another, which results in no change to an amino acid or one amino acid being changed to another. This may result in a non-functional protein or have little effect on the protein. Sickle cell disease is caused by a substitution missense mutation.

DNA base sequence

DNA base sequence	CCG	GTG	TCA	TGT	GGT	AAG
mRNA codons	GGC	CAC	AGU	ACA	CCA	UUC
resulting amino acid sequence	glycine	histidine	serine	threonine	proline	phenylalanine

Effect of substitution

		A replaces T here ↓				
Mutated DNA base sequence	CCG	GAG	TCA	TGT	GGT	AAG
mRNA codons	GGC	CUC	AGU	ACA	CCA	UUC
resulting amino acid sequence	glycine	leucine	serine	threonine	proline	phenylalanine

Substitution – nonsense

Nonsense mutations result in a premature stop codon being produced that results in a shorter protein. Duchenne muscular dystrophy is caused by a substitution nonsense mutation.

Substitution – splice-site

A substitution of one base for another at a splice-site results in some introns being retained and/or some exons not being included in the mature transcript. Beta thalassemia is a disease that is caused by a substitution splice-site mutation.

TOP TIP

Note: a splice-site is the boundary of an exon and an intron. A base change can disrupt RNA splicing, resulting in the loss of exons or inclusion of introns and an altered protein coding sequence.

Insertion and deletion mutations

These single gene mutations result in frame-shift mutations. A frame-shift mutation causes all of the codons and all of the amino acids after the mutation to be changed. This has a major effect on the protein produced. An example of a frame-shift insertion is found in the genome of those with Tay-Sachs disease and a frame-shift deletion is found in the genome of those with cystic fibrosis.

Effect of deletion

	G deleted here ↓		all bases moved one place to the left			
mutated DNA base sequence	CCG	TGT	CAT	GTG	GTA	AG
mRNA codons	GGC	ACA	GUA	CAC	CAU	UC
resulting amino acid sequence	glycine	threonine	valine	histidine	histidine	

Effect of insertion

	additional G added here ↓		all bases moved one place to the right				
mutated DNA base sequence	CCG	GGT	GTC	ATG	TGG	TAA	G
mRNA codons	GGC	CCA	CAG	UAC	ACC	AUU	C
resulting amino acid sequence	glycine	proline	glutamine	tyrosine	threonine	isoleucine	

TOP TIP

Remember:

1) Mention codons and base sequences when asked about changes to the structure of a gene.
2) Mention amino acids when asked about changes to the structure of a protein.

TOP TIP

Remember types of single gene mutations as 'SID':

- Substitution
- Insertion
- Deletion.

Quick Test 10

1. State what the genome is made up of.
2. State the function of non-protein coding sequences.
3. State what is meant by the term 'single gene mutation'.
4. Explain the term 'frame-shift mutation'.
5. In what way does a 'splice-site mutation' affect the mature mRNA transcript?

Chromosome structure mutations

Change in chromosome number

Mutations vary in size from those that affect a single nucleotide base pair in one gene to mutations that affect many genes on a chromosome.

Some mutations alter chromosome number in every body cell. These mutations are caused by a gamete containing more or less than 23 chromosomes. For example, Down's syndrome can be caused by a sperm or egg containing an extra copy of chromosome number 21. The resulting zygote contains 47 instead of 46 chromosomes.

Change in chromosome structure

If whole sections of chromosomes containing many genes are broken, rearranged or lost, this is a mutation of chromosome structure. The substantial changes in chromosome mutations often make them lethal.

- **Deletion** – where a section of a chromosome is removed. This occurs when two breaks occur along the length of the chromosome, and the middle segment of chromosome containing many genes is lost. The broken ends of the remaining chromosome join together.
- **Duplication** – where a section of a chromosome is added from its homologous partner. This occurs when a broken segment from a similar neighbouring chromosome is inserted, duplicating a specific set of genes.
- **Inversion** – where a section of chromosome is reversed. This is when two breaks occur along the length of a chromosome. The segment of chromosome between the breaks rotates 180 degrees and reattaches, reversing the gene sequence.
- **Translocation** – where a section of a chromosome is added to a chromosome, not its homologous partner. This occurs when a segment breaks off the end of one chromosome and attaches to the end of a neighbouring chromosome, adding additional genes.

TOP TIP

Remember 'DDTI' for types of chromosome mutation:

- **D**eletion
- **D**uplication
- **T**ranslocation
- **I**nversion.

TOP TIP

Remember that chromosomes are arranged in pairs. An **homologous** pair contains one chromosome from each parent. Both chromosomes in a pair are the same length and contain genes at similar locations.

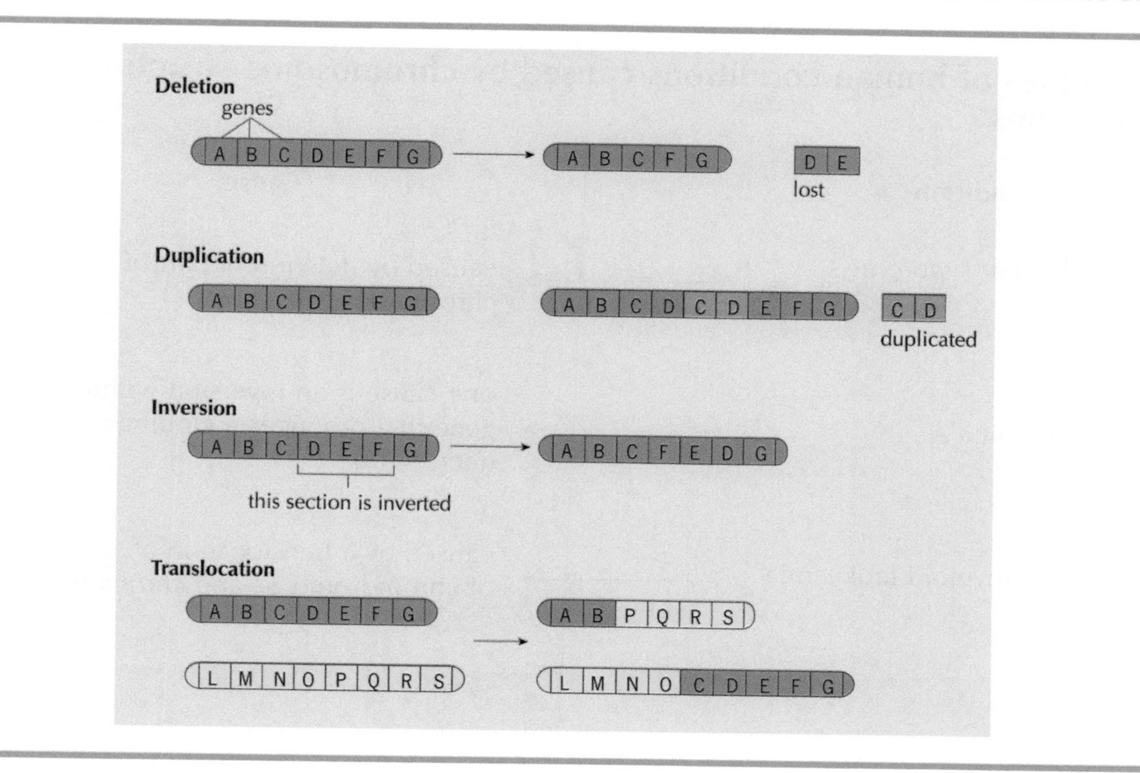

Importance of mutations and gene duplication in evolution

Mutations can be harmful or beneficial. Duplication of genes allows potential for beneficial mutations to occur in a duplicated gene. This is because the second copy of the gene has the potential to undergo random mutation itself and may produce a protein that could confer a survival advantage. The original gene is still expressed to produce its protein.

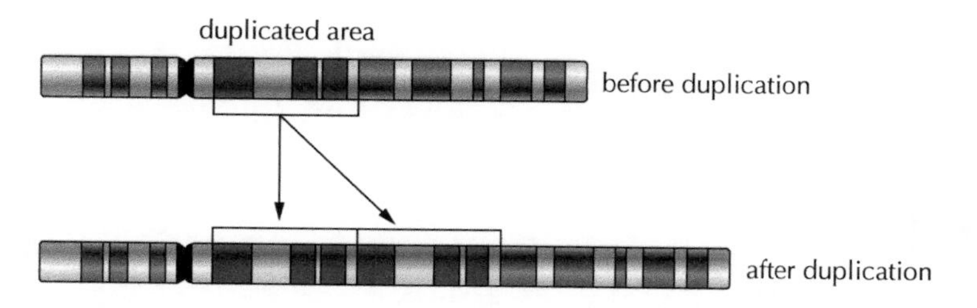

GOT IT? ☐ ☐ ☐

Examples of human conditions caused by chromosome structure mutations

Condition		Cause
Cri-du-chat syndrome	→	caused by deletion of part of chromosome 5
Haemophilia A	→	one cause is an inversion within the gene that produces a clotting factor (factor VIII)
Chronic myeloid leukaemia	→	caused by a translocation of sections of chromosome 22 and chromosome 9

Quick Test 11

1. State the name of a genetic condition in humans caused by a change in chromosome number.
2. Name the four types of chromosome structure mutations.
3. State the type of chromosome structure mutation that results in a segment of one chromosome attaching to the end of a neighbouring chromosome.
4. Explain the importance of duplication in evolution.

Evolution

Evolution is the changes in organisms over generations as a result of genomic variation.

Inheritance

Genomic material can be inherited in two ways:

1. **Vertical gene transfer**.
2. **Horizontal gene transfer**.

Vertical gene transfer occurs when genes are transferred from parent to offspring in the next generation as a result of sexual or asexual reproduction.

Horizontal gene transfer is when genes are transferred between individuals in the same generation.

TOP TIP

Remember: faster evolutionary change occurs in prokaryotes due to horizontal gene transfer within the same generation.

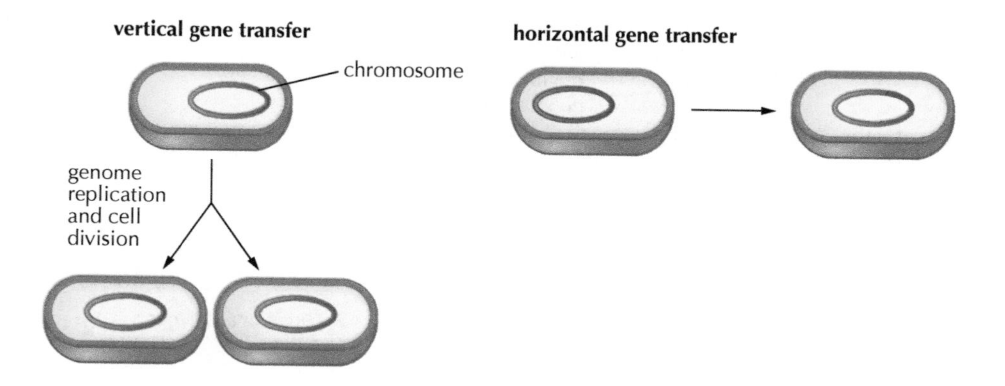

Natural selection

Natural selection is the non-random increase in frequency of DNA sequences that increase survival and the non-random reduction in the frequency of deleterious (potentially harmful) sequences.

Darwin's Theory of Natural Selection:

1. Organisms produce more offspring than the environment can support.
2. Genetic variation occurs within individuals of a population.
3. Individuals compete for available resources such as food and mates. Natural selection occurs when there are selection pressures.
4. The best adapted individuals in a population survive to reproduce, passing on the favourable alleles that confer a selective advantage.
5. These alleles increase in frequency within the population.

Selection

Natural selection can affect the phenotype frequency of a measureable trait, such as height or mass, within a large population in three ways.

1. **Stabilising selection** occurs within a stable environment and selects for an average phenotype within a population and extremes of the phenotype are selected against.
2. **Directional selection** occurs within a changing environment and one extreme of the phenotype is selected for.
3. **Disruptive selection** occurs when two different types of environment or resources become available. Two or more phenotypes are selected for.

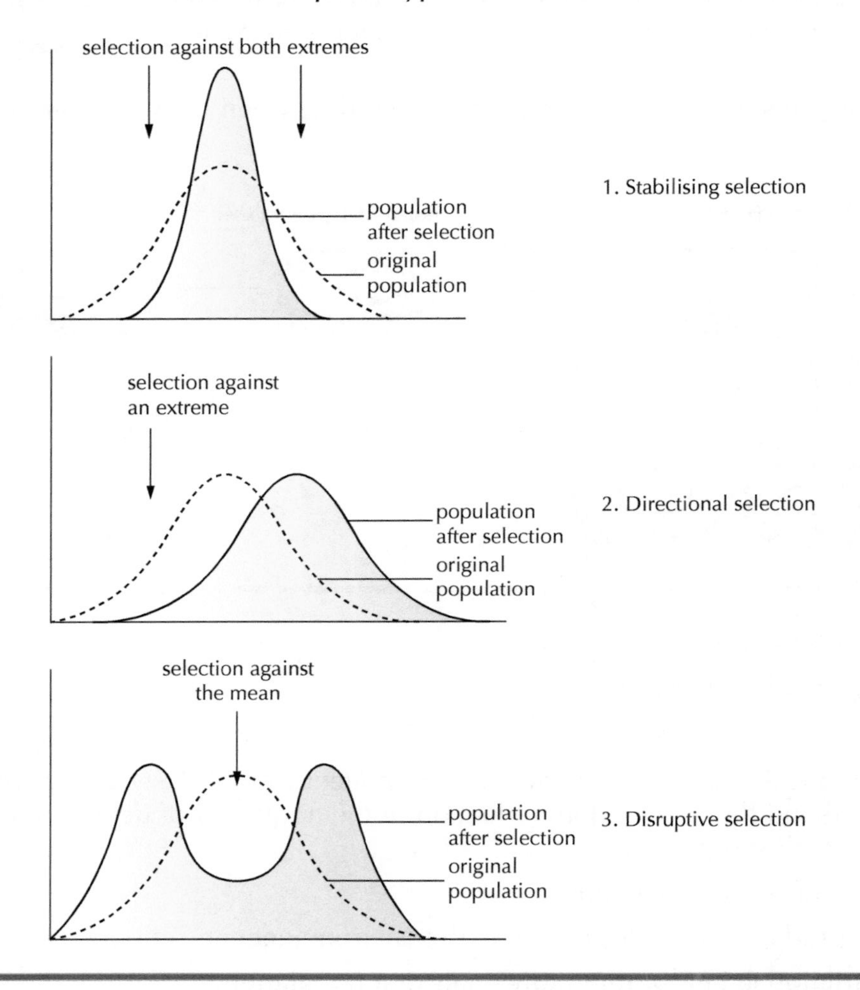

Speciation

The generation of a new biological **species** as a result of isolation, mutation and selection is called **speciation**.

1. **Allopatric speciation** – a population is split into two sub-populations, A and B, by a geographical barrier such as a river which prevents gene flow. Mutations occur separately in populations A and B, followed by natural selection over many generations, resulting in two new species, A and B, which can now no longer interbreed.
2. **Sympatric speciation** – a population is split into two-sub populations, A and B, by behavioural or ecological barriers which prevent gene flow. Different mutations occur in each sub-population. Natural selection acts differently in both sub-populations. Over many generations each sub-population becomes so genetically different they can no longer interbreed and produce fertile offspring. Speciation has occurred.

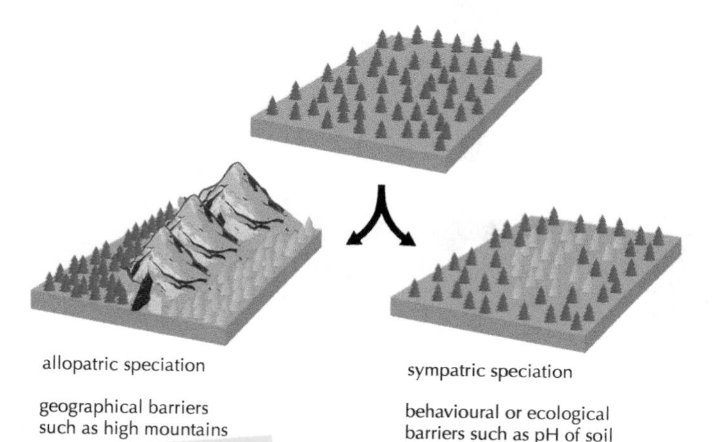

TOP TIP

Remember: a species is a group of organisms capable of interbreeding and producing fertile offspring and which does not normally breed with other groups.

Quick Test 12

1. Give one advantage to prokaryotes of horizonatal gene transfer compared to vertical gene transfer in eukaryotes.
2. Suggest why some individuals within a population might fail to survive and reproduce.
3. Describe vertical gene transfer in eukaryotes.
4. Explain the importance of an isolating barrier in speciation.
5. Which types of barriers are involved in sympatric speciation?

Genomic sequencing

The analysis of the sequence of bases within an organism's DNA is called genomic sequencing, and is carried out within a laboratory. This process involves **bioinformatics**, computer technology that stores, organises and statistically analyses sequencing information.

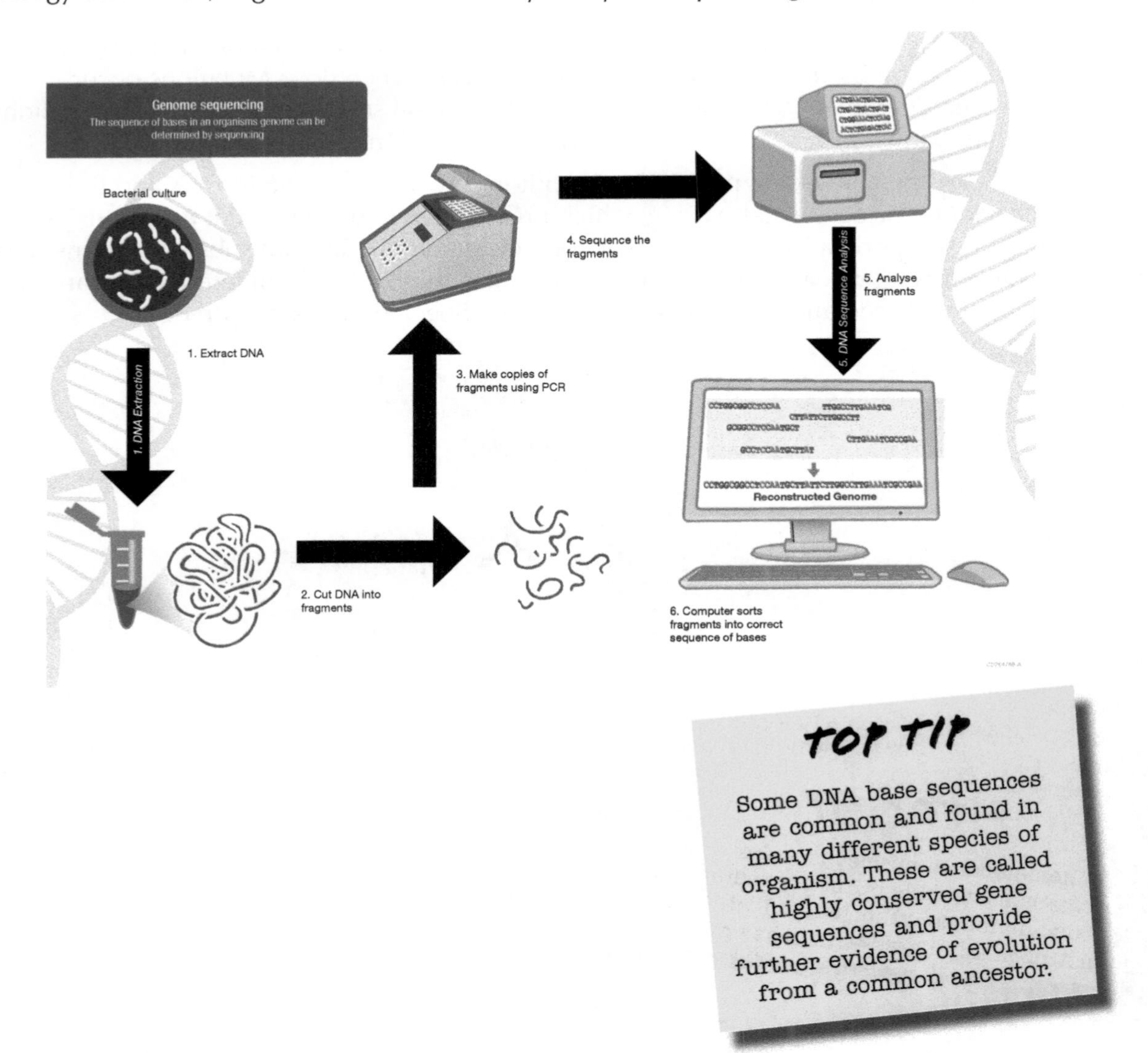

TOP TIP

Some DNA base sequences are common and found in many different species of organism. These are called highly conserved gene sequences and provide further evidence of evolution from a common ancestor.

Genomic sequencing

The genomic sequence of nucleotide bases can be determined for individual genes and **genomes**. Genomics is the study of the entire genome of an individual or a species. It is very useful to know the exact sequence of nucleotide bases as it allows analysis and comparison of the entire DNA sequence between one organism and another.

There are approximately 3.2 billion base pairs in the human genome alone. In order to identify sequences similar to known genes, computer programs are needed. In fact, the new science of bioinformatics combines computer and statistical analysis in order to compare sequence data between individuals and between species.

The human genome was completed on the 14th of April 2003. Many other genomes have been sequenced since then. These include:

Genome sequenced		
Disease-causing organism	Pest species	Model organism
Plasmodium falciparum A parasite that causes malaria 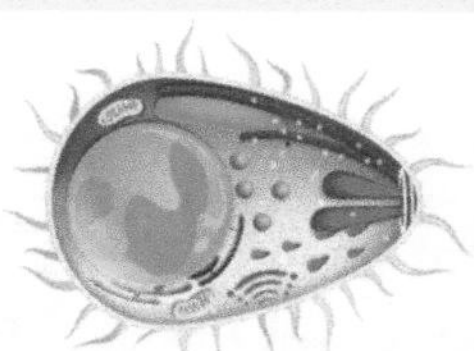	 Mountain pine beetle  Locust	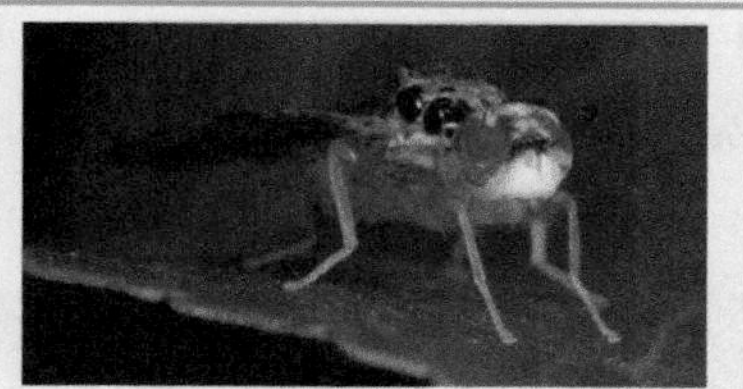 Mediterranean fruit fly 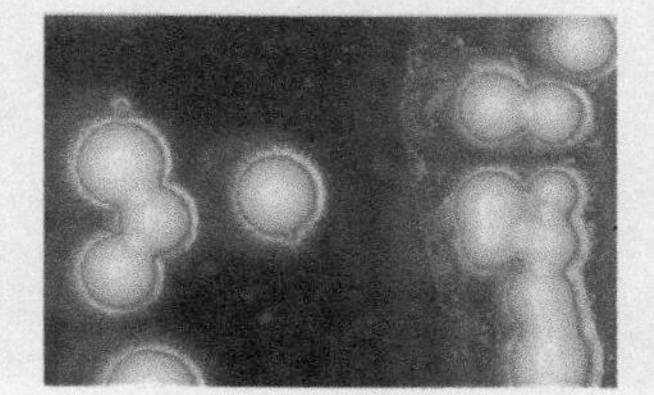 Yeast  Mouse

TOP TIP

Remember that a **model organism** is a non-human species that is easy to grow and keep in a laboratory and whose biology has been widely studied.

TOP TIP

Sequencing the genomes of model organisms, pest species and disease-causing organisms helps scientists to learn how to help fight disease, increase food security and understand cellular processes.

Comparing genomes

Comparison of genomes helps scientists identify common genes between organisms and also helps identify genes that make organisms unique.

Genome size does not correlate with complexity; for example, the marbled lungfish has the largest recorded genome of any eukaryote (132.8 billion base pairs).

Comparison of human and yeast genomes			
Organism	Estimated size of genome (base pairs)	Chromosome number	Estimated gene number
Human	3.2 billion	46	20,000
Yeast	12 million	16	6000

The three domains of life

Comparison of sequences provides evidence of the three domains of life:

1. Bacteria (prokaryotes).
2. Archaea (prokaryotes living in extreme conditions of heat or salinity).
3. Eukaryotes (plants, animals and fungi).

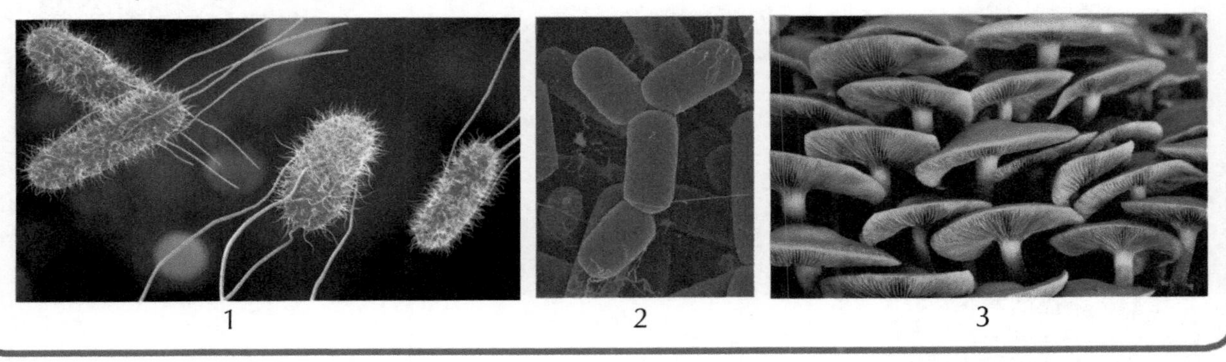

1 2 3

Sequence of events in evolution

The sequence of events in the evolution of life can be determined by scientists using:

1. fossil evidence
2. sequence data.

A combination of both of the above support the theory that all life evolved over millions of years from a universal ancestor. Changes in gene sequences can be used to study the evolutionary relatedness among groups of organisms combined with, if available, fossil evidence. Fossil evidence adds timescales to evolutionary events.

Evolutionary events starting with the earliest
1. cells
2. last universal ancestor
3. prokaryotes
4. photosynthetic organisms
5. eukaryotes
6. multicellularity
7. animals
8. vertebrates
9. land plants

Phylogenetics

Phylogenetics is the study of evolutionary history and relationships between organisms. The most common way of illustrating how organisms are related is to use a diagram called a phylogenetic tree. Each branch in the tree represents the evolution of a distinct species over time from a common ancestor. The more closely related two species are, the more recently they shared a common ancestor.

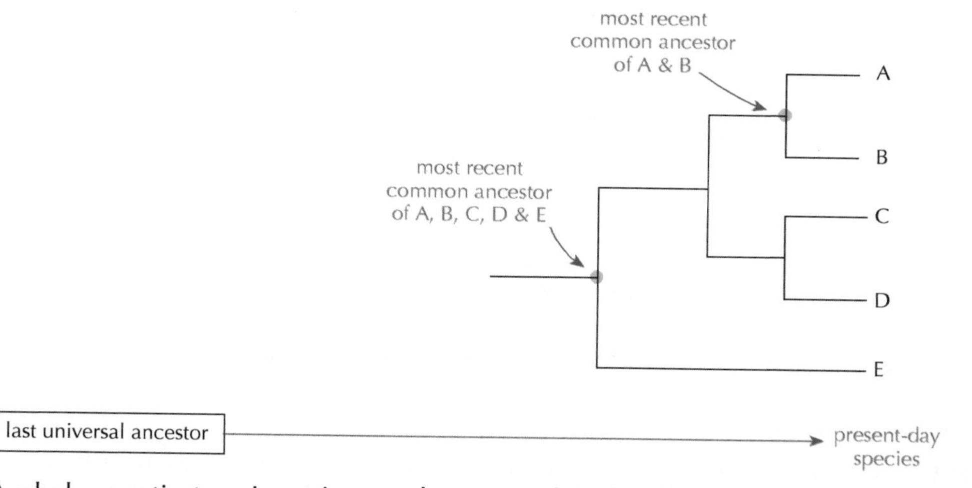

A phylogenetic tree is an incomplete record as fossil evidence and DNA sequence data are not available for every organism that ever lived. However, phylogenetic trees are useful to see when different lineages (shared evolutionary pathways) and sequences diverged (split). As a rule, the most recently diverged organisms on a tree are the most closely related. The point of divergence is where two species shared a common ancestor.

Molecular clocks

Gene sequences mutate over time. If two species diverged a long time ago, they will have many differences in their sequence data for a particular protein. If they diverged recently, they will have fewer differences in their sequence data for the same protein.

Some sequences of DNA (and the amino acids they code for) mutate at a constant rate. It is these sequences that are used as molecular clocks. The differences in gene sequence over time (combined with fossil evidence of divergence) can be used to estimate when two species diverged from a common ancestor.

Molecular clocks assume a constant mutation rate and show differences in DNA sequence or amino acid sequence due to mutation over time.

GOT IT? ☐ ☐ ☐

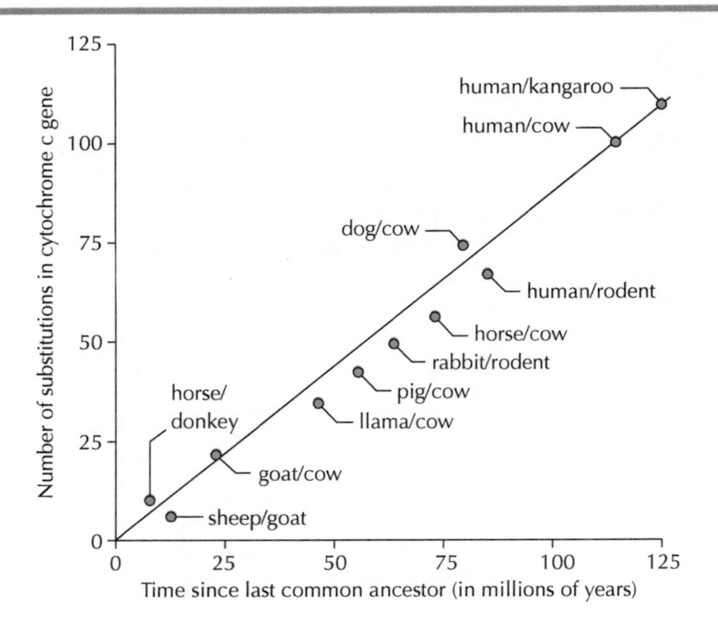

A molecular clock showing evolutionary relatedness of different species.

In the molecular clock diagram above:

Rabbit and rodent species have 50 substitution mutation differences in their cytochrome c gene sequence. Human and cows have 100 substitution mutation differences in their cytochrome c gene sequences. As shown in the diagram above, rabbits and rodents have a more recent common ancestor than humans and cows.

Pharmacogenetics and personalised medicine

An individual's personal genome sequence can be:

1. analysed to predict the likelihood of developing certain diseases
2. used to select the most effective drugs and dosage to treat their disease – this is known as **personalised medicine**.

Pharmacogenetics is the use of genome information in the choice of drugs.

Quick Test 13

1. Why it is useful to know the exact sequence of nucleotide bases in organisms?
2. What does 'bioinformatics' involve?
3. State the two types of evidence needed to create a phylogenetic tree.
4. State the property a sequence of DNA must have in order to be used in a molecular clock.
5. State the names of the three domains of life.

Metabolic pathways

Chains of different enzyme-catalysed biochemical reactions called **metabolic pathways** make up the **metabolism** of a cell.

Reversible and irreversible steps

Each step of a metabolic pathway is controlled by an enzyme. Some steps are reversible, meaning that metabolite A may be converted to metabolite B, and metabolite B can be converted back again to metabolite A and used perhaps in an alternative pathway. The cell has control over the pathway.

metabolite A → metabolite B → metabolite C
↑ enzyme 1 ↑ enzyme 2

An irreversible step results in one metabolite being fully converted to another, with no alternative pathway available.

Alternative routes

Specific steps in a pathway can be bypassed using an alternative route or 'short cut', so that metabolite A is converted directly to metabolite C, for example.

Anabolic and catabolic

Anabolic pathways make or synthesise new molecules from basic building blocks using energy in the form of adenosine triphosphate (ATP); for example, building up protein molecules from amino acids.

Catabolic pathways break down large complex molecules into their smaller subunits, with the release of energy (ATP). For example, the process of cell respiration involves the breakdown of glucose with oxygen to form ATP, carbon dioxide and water.

TOP TIP

Anabolic reactions – energy IN
Catabolic reactions – energy OUT
(Remember: put out the cat!)

Integrated metabolic pathways

Anabolic and catabolic pathways can become interdependent and form an integrated metabolic pathway. The catabolic pathway of cell respiration producing ATP is linked to the anabolic pathway of synthesis of proteins from amino acids, which requires energy.

GOT IT?

Functions of membrane proteins

Molecules are needed for reactions. In order for these to enter cells or be produced, the following two types of membrane protein are needed:

1. **Protein pores** are 'channels' in large protein molecules that span the plasma membrane. They act as 'gateways' for large molecules to pass through by diffusion.

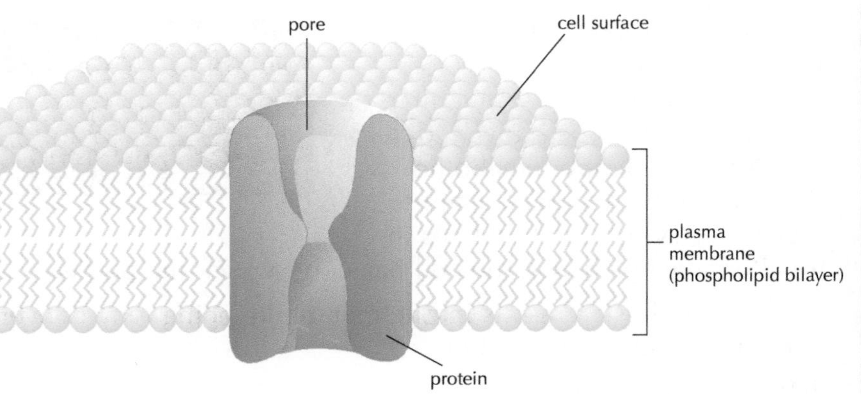

TOP TIP

Remember, to maintain cellular metabolism both optimal enzyme activity and high diffusion rates are necessary.

2. **Protein pump** is a carrier protein on the plasma membrane that moves (pumps) molecules and ions across the membrane from a low concentration to a high concentration (the opposite of diffusion). The movement of the carrier protein requires energy in the form of ATP. This is an active process.

   Like many enzymes, protein pores and pumps are embedded in the cell membrane.

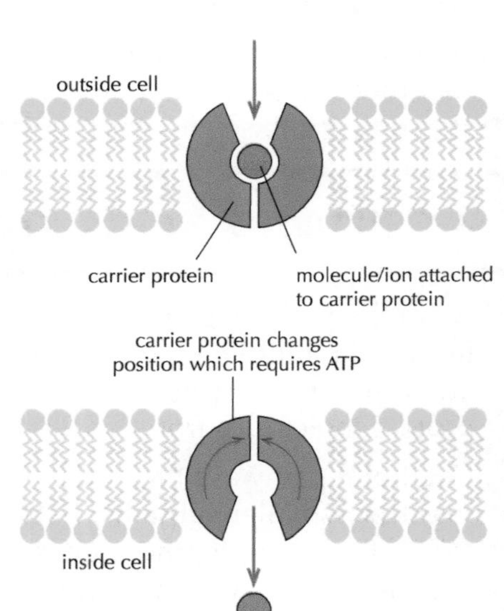

Quick Test 14

1. State two different types of metabolic pathway.
2. Explain the advantage to a cell of having a reversible step within a metabolic pathway.
3. Explain the function of a protein pump.
4. State the function of protein pores in the plasma membrane.
5. State one method of molecular transport that involves a carrier protein.

Control of metabolic pathways

Each step within a metabolic pathway is controlled by the presence of one specific enzyme, coded for by one gene. If an enzyme within the pathway is missing, or does not work properly due to mutation, the chain of chemical reactions is disrupted.

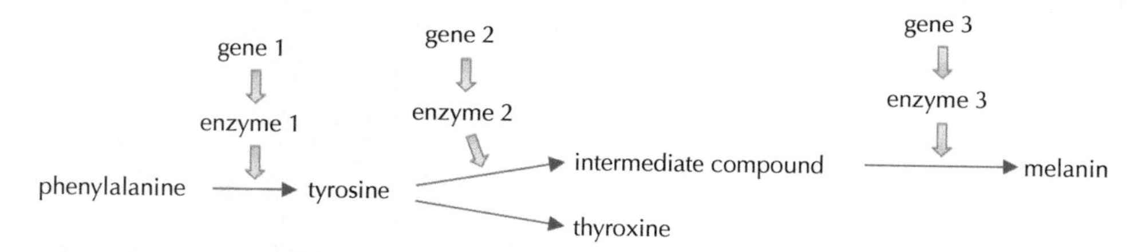

Enzyme action

The minimum energy required for chemical molecules to be able to react together within a cell is called **activation energy**. When the reactants have reached a threshold energy level, at the **transition state**, the reaction can proceed and products are formed. Enzymes are catalysts that lower activation energy, speeding up chemical reactions that would otherwise be too slow to sustain life.

TOP TIP

Enzymes speed up chemical reactions within the cell, but do not take any part in the reaction, remaining unchanged.

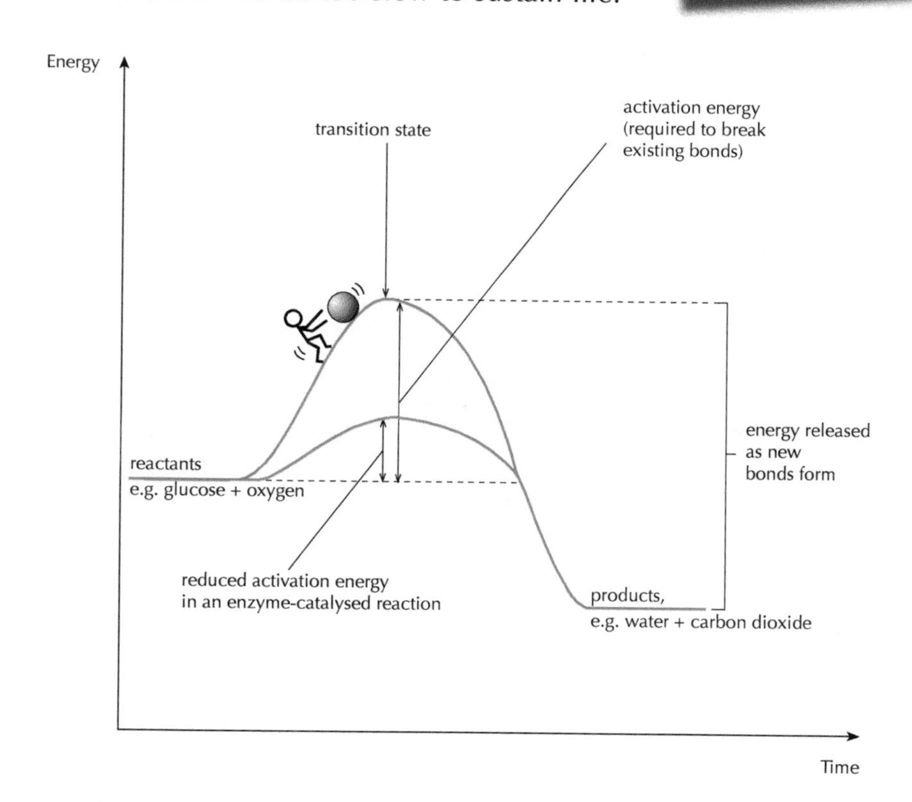

Enzyme structure

Enzymes are three-dimensional globular protein molecules that are sensitive to both pH and temperature. The specific part of the enzyme molecule that connects with a substrate molecule is called the **active site**.

The shape of the active site on an enzyme molecule is specific to only one substrate. If the shape of a substrate molecule is complementary to the active site of the enzyme, it will have a high affinity for the active site. Products of enzyme substrate reactions have a low affinity for the active site of the enzyme.

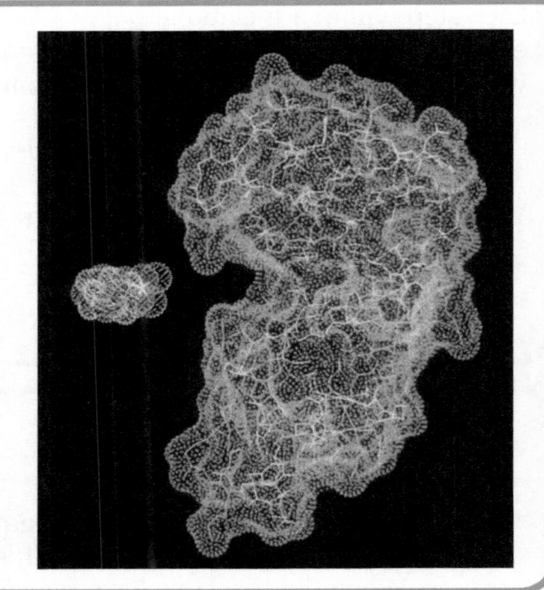

TOP TIP

The active site of an enzyme is complementary to the shape of a specific substrate molecule. Imagine two jigsaw pieces of the correct complementary shapes being able to fit together.

Induced fit

Substrate molecules that have a similar shape but not exactly complementary to the active site of a specific enzyme may still have some affinity and be able to react. The active site changes shape to better fit the substrate after the substrate binds. This is called **induced fit**.

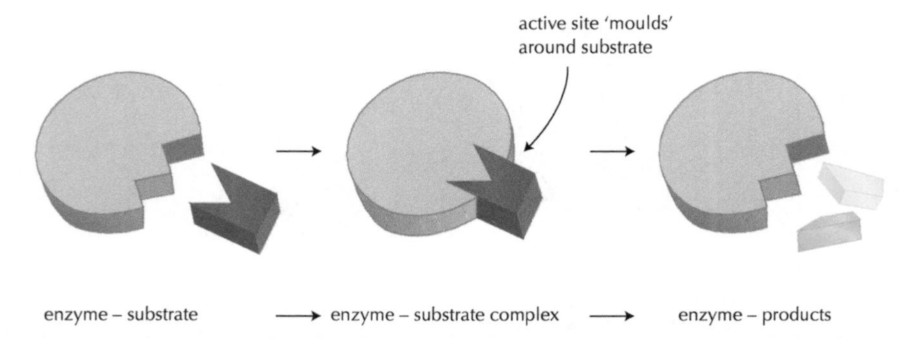

Orientation of active site and substrate molecules

The shape of the active site determines the position or **orientation** of the two substrate molecules. This ensures the reactants are in the best possible position for the reaction to proceed. This also reduces the activation energy. The products have a low affinity for the active site and this allows them to be released.

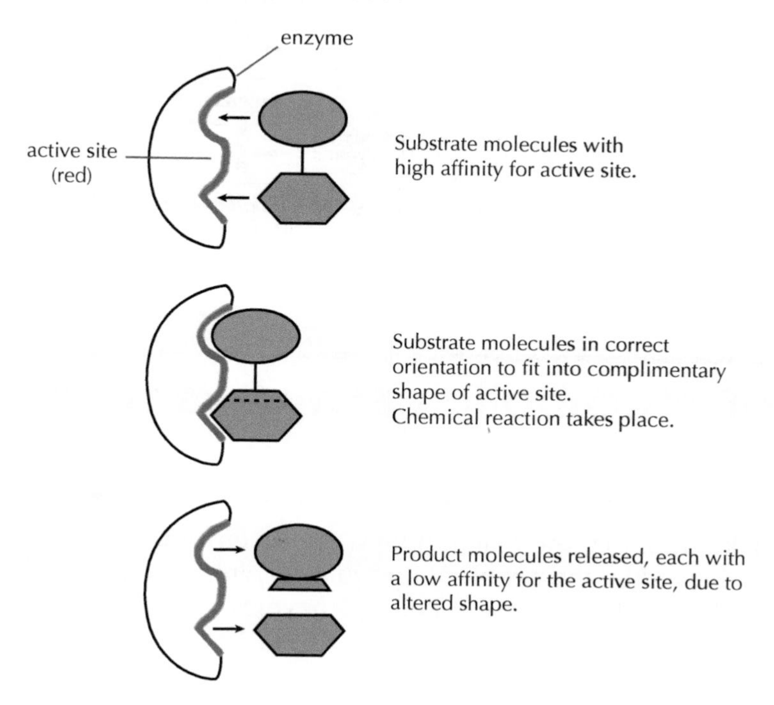

Factors affecting enzyme action

The direction and rate of an enzyme-catalysed reaction are influenced by both the concentration of the substrate and the product(s).

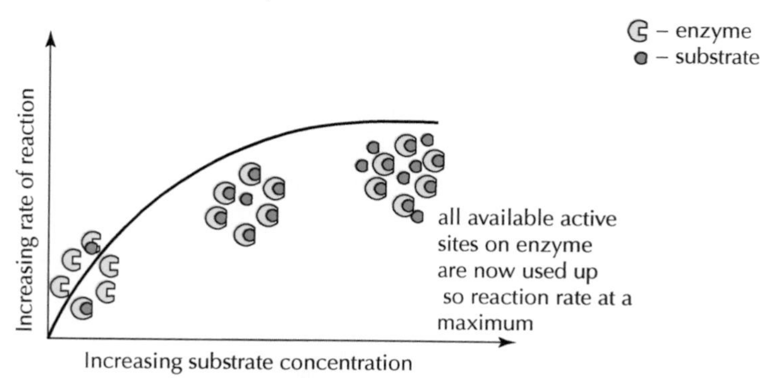

Another way an enzyme-catalysed reaction can be regulated occurs when the product of the last reaction in a metabolic pathway reaches a critical concentration and then inhibits the enzyme that catalyses the first reaction of the pathway. This is an example of **feedback inhibition**.

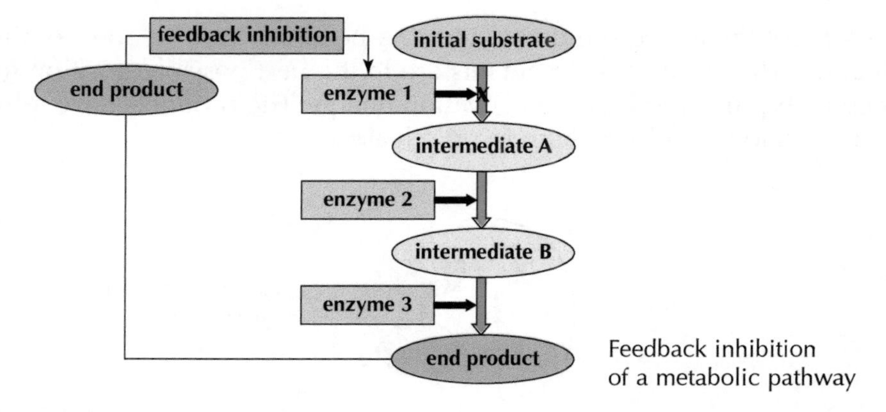

Feedback inhibition of a metabolic pathway

Control of metabolic pathways through inhibitors

As well as feedback inhibition, the reaction between an enzyme and its substrate molecules can be affected by the action of an **inhibitor**. Inhibitor molecules can slow down metabolic pathways that are controlled by enzymes.

There are two types of inhibitor molecule:

1. **Competitive inhibitor.**
2. **Non-competitive inhibitor.**

Competitive inhibitors

Competitive inhibitor molecules slow down the rate of reaction by actively competing with substrate molecules for the active site of an enzyme. If the active site contains an inhibitor molecule, the substrate molecule cannot enter.

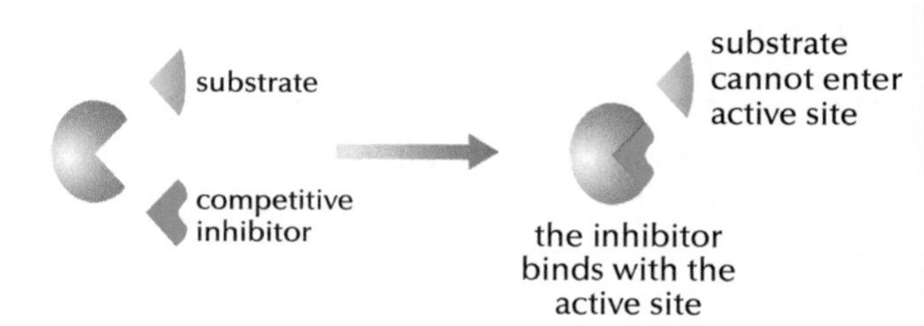

The effect of the inhibitor can be reduced by increasing the concentration (number) of substrate molecules, which increases the reaction rate. Substrate molecules outnumber inhibitor molecules and are more likely to win the race for the active sites!

Increasing the concentration of enzyme (number of active sites) can also reduce the effect of a competitive inhibitor. Competitive inhibition is therefore reversible.

Non-competitive inhibitors

Non-competitive inhibitor molecules bind away from the active site. Once in position, the inhibitor changes the three-dimensional structure of the active site, preventing connection with substrate molecules, and decreasing the rate of reaction. Altering the enzyme or substrate concentration has no effect.

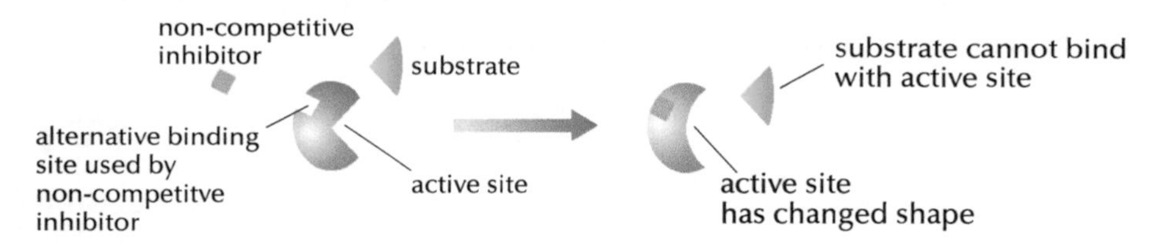

Quick Test 15

1. Describe how lowering the activation energy affects the rate of a chemical reaction.
2. Describe the principle of induced fit in an enzyme substrate reaction.
3. Describe the effect of increasing substrate concentration on the rate of reaction.
4. Describe the effect of orientation of substrate molecules in the active site of an enzyme on the resulting enzyme substrate reaction.
5. State the effect of an inhibitor molecule on the rate of reaction between an enzyme and substrate.
6. State the type of inhibitor that alters the shape of the active site of an enzyme.
7. Describe how the effects of a competitive inhibitor can be reduced.

Cellular respiration

The process of life

Cell respiration occurs in all living cells and consists of a series of reactions forming a metabolic pathway that provides energy. The main respiratory substrate is glucose, which is broken down, hydrogen ions and electrons are removed by **dehydrogenase** enzymes and ATP is released.

Adenosine triphosphate (ATP)

ATP is a high-energy chemical compound needed by cells to do 'work', such as DNA replication, cell division, active transport, synthesis of new molecules and muscle contraction.

Simplified diagram of ATP structure

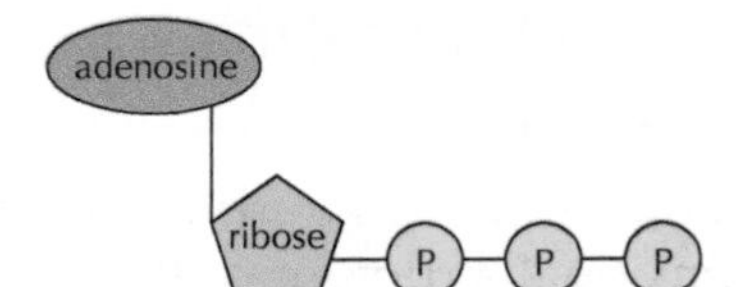

Phosphorylation is a reaction that involves the addition of a phosphate to a molecule. The source of the phosphate is usually ATP. Phosphorylation increases the energy content of the molecule to which the phosphate group is attached.

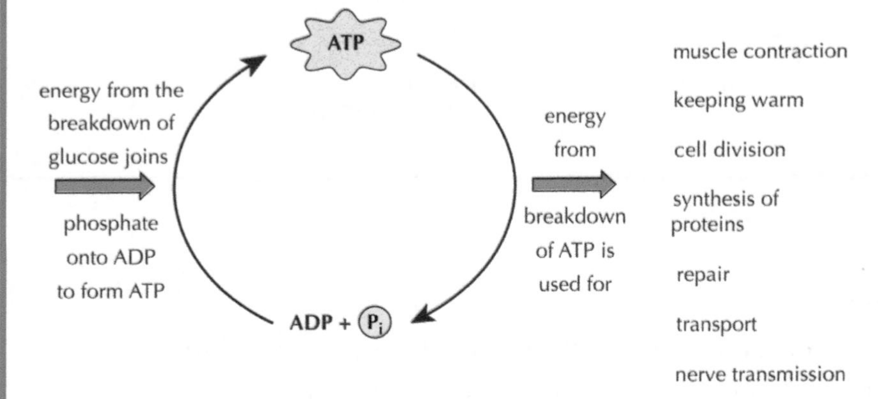

TOP TIP

A cell cannot use energy directly from the breakdown of glucose; it must be in the form of ATP.

TOP TIP

There is only around 50 g of ATP in the body at any one time, as it is constantly being broken down and regenerated.

Metabolic pathways of cellular respiration

Cell respiration is the breakdown (oxidation by the removal of hydrogen ions and electrons) of glucose molecules within a cell to release energy for the synthesis of ATP.

There are three pathways in the process of cellular respiration:

1. **Glycolysis,** which occurs in the cytoplasm.
2. **Citric acid cycle,** which occurs in the central matrix of a mitochondrion.
3. **Electron transport chain** which occurs on the inner mitochondrial membrane.

Glycolysis

Glycolysis takes place in the cytoplasm of a cell when glucose is broken down to form **pyruvate** and does not require oxygen. Initially, 2 ATP molecules are used up as an energy investment phase. This is followed by reactions that form 2 pyruvate molecules, which is an energy pay-off phase because 4 ATP molecules are produced. This is a net gain of 2ATP overall.

Hydrogen ions and electrons are removed from the intermediates by a dehydrogenase enzyme during the pay-off phase. Hydrogen ions are extremely unstable and are immediately picked up by the coenzyme NAD to form NADH.

TOP TIP

Glycolysis: 'glyco' – means 'glucose', 'lysis' – means 'to break'.

TOP TIP

Remember that glucose and intermediates are phosphorylated during glycolysis.

Quick Test 16

1. State three examples of functions carried out by a cell that require energy in the form of ATP.
2. Write an equation to illustrate the process of phosphorylation.
3. Explain why there is only a very small mass of ATP present in the body at any one time.
4. State one example of a hydrogen carrier in cellular respiration.
5. State the name of the molecule that glucose is finally broken down to in glycolysis.
6. State where glycolysis takes place in a cell.
7. Explain why a cell receives a net gain of 2 ATP during glycolysis.

Citric acid cycle

The citric acid cycle takes place in the matrix of a mitochondrion, and occurs only if a cell has oxygen. Pyruvate, produced in glycolysis, is broken down into **acetyl coenzyme A**, with the release of carbon dioxide and more hydrogen ions and electrons picked up by NAD. The acetyl group of acetyl coenzyme A combines with **oxaloacetate** to form **citrate**.

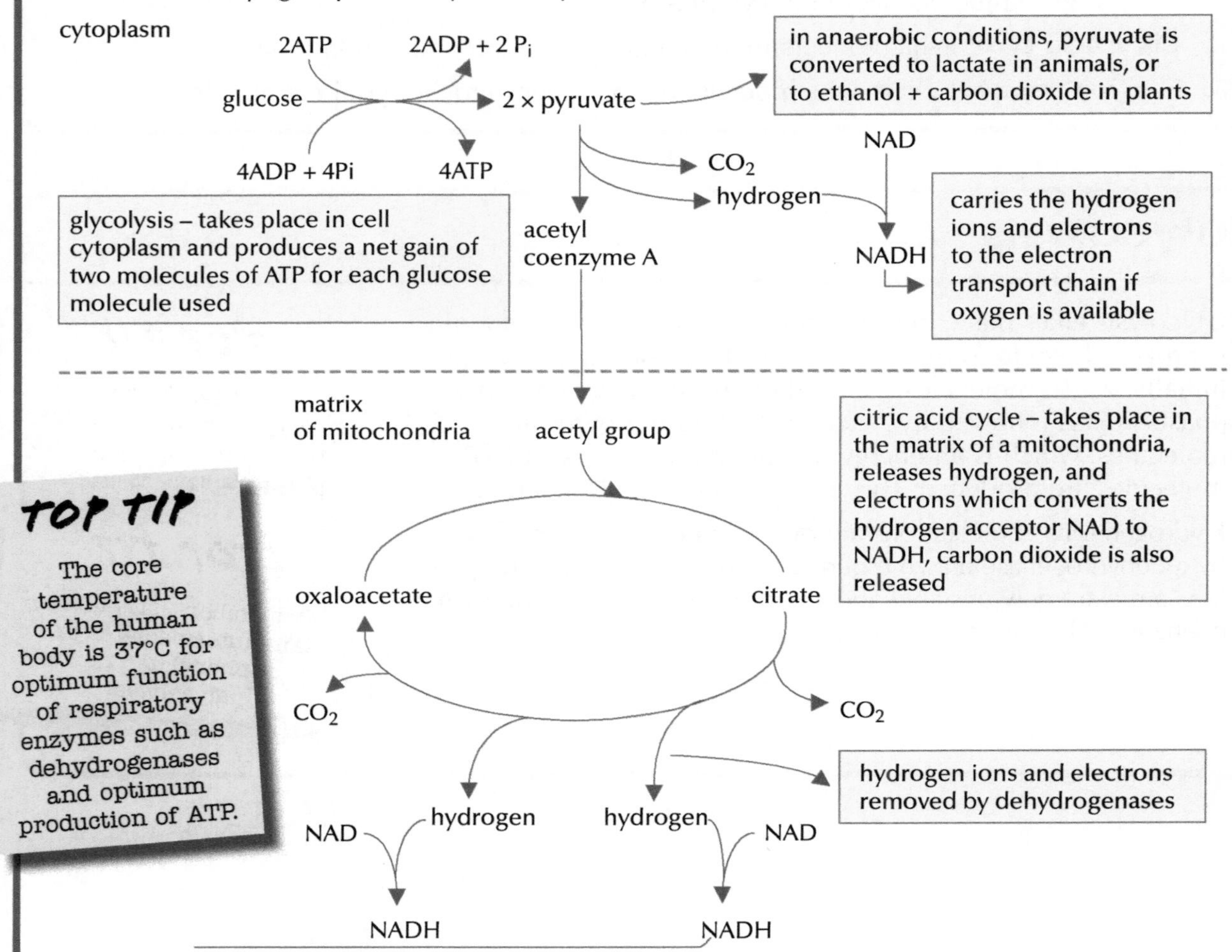

TOP TIP

The core temperature of the human body is 37°C for optimum function of respiratory enzymes such as dehydrogenases and optimum production of ATP.

- Pyruvate (from glycolysis) is broken down to an acetyl group that combines with coenzyme A to form acetyl coenzyme A (acetyl coA).
- Dehydrogenase enzymes remove hydrogen ions and electrons and pass them onto to the coenzyme NAD to form NADH.
- The acetyl group from acetyl coenzyme A combines with oxalacetate to form citrate.
- Several enzyme-controlled steps occur within the cycle, resulting in the regeneration of more oxaloacetate.
- Carbon dioxide is released, ATP synthesised and NADH generated.

Electron transport chain

The electron transport chain takes place on the inner membrane of a mitochondrion. Here, a chain of carrier proteins accept hydrogen ions and electrons from NADH. The electrons and hydrogen ions are finally transferred to oxygen to form water.

ATP synthesis

The sythesis of ATP is brought about by the action of an enzyme called **ATP synthase**, which sits in the inner membrane of a mitochondrion. It catalyses the synthesis of ATP from ADP and P_i.

- Electrons flow along the chain of protein electron carrier molecules.
- Energy released from the flow of electrons is used to pump hydrogen ions across the inner membrane of the mitochondria.
- The return flow of hydrogen ions rotates part of a membrane protein called ATP synthase, phosphorylating ADP + P_i to ATP.
- When the hydrogen ions and electrons reach the end of the protein carrier chain, they combine with oxygen to form water.

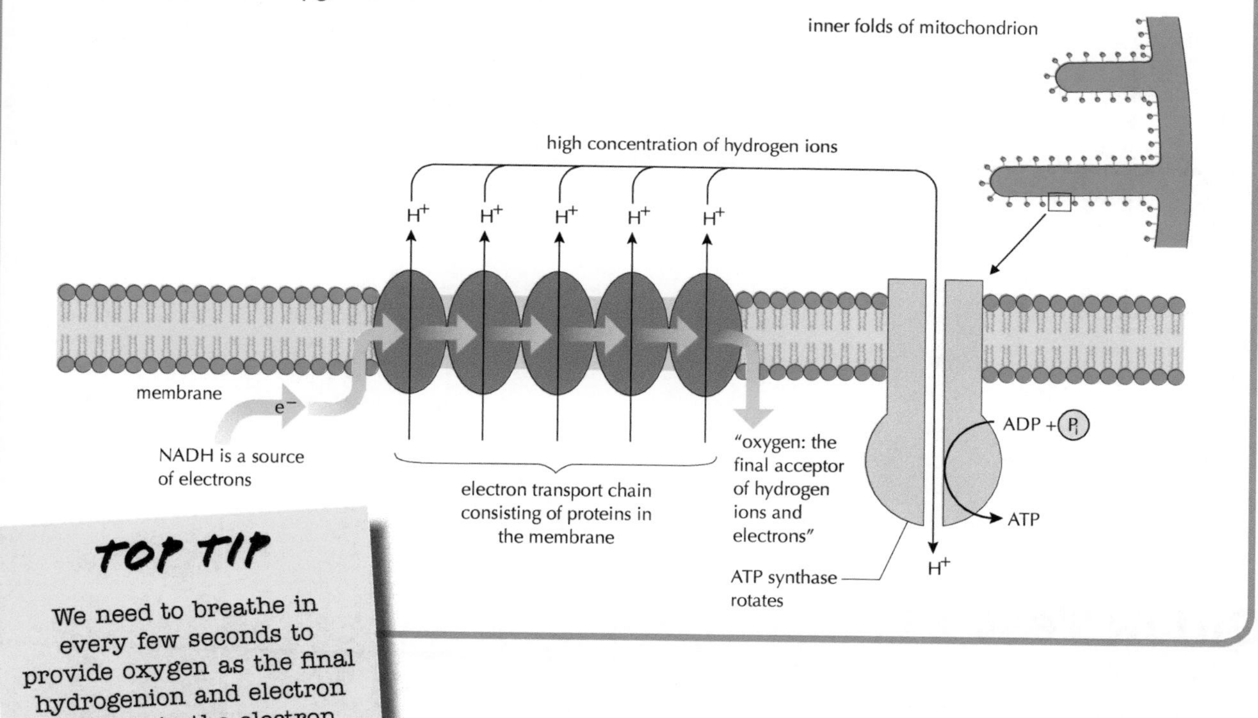

TOP TIP

We need to breathe in every few seconds to provide oxygen as the final hydrogenion and electron acceptor in the electron transport chain.

Fermentation

When oxygen is unavailable, **fermentation** takes place in the cytoplasm. This is much less efficient than aerobic respiration. The pyruvate is converted into **lactate** in animal cells and **ethanol** and carbon dioxide in yeast and plant cells. Only 2 ATP are produced.

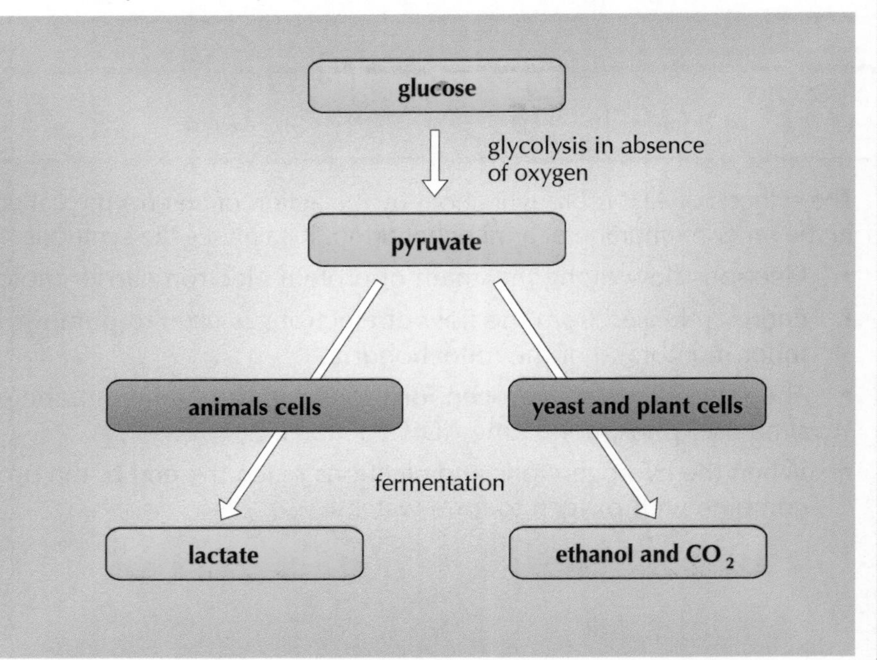

TOP TIP

In animals, fermentation is reversible, but in yeast and plant cells, it is irreversible.

Quick Test 17

1. State where in a cell the citric acid cycle takes place.
2. State the hydrogen carrier in the citric acid cycle.
3. Which stage of cell respiration involves a chain of protein carrier molecules?
4. Describe the role of ATP synthase in the production of ATP molecules.
5. State the final hydrogen acceptor in the electron transport chain.

Metabolic rate

The **metabolic rate** of a living organism refers to the speed with which energy is used up to survive.

Comparison of metabolic rates between different organisms

The energy to drive metabolism comes from the breakdown of glucose during aerobic respiration. Because oxygen is used up and carbon dioxide and heat energy released, it is possible to compare the metabolic rates between different organisms by looking at how much:

1. oxygen is consumed
2. carbon dioxide is produced
3. energy is released in the form of heat.

TOP TIP

The heat energy released during metabolism can be measured using a calorimeter.

The oxygen consumption per hour of a small organism can be measured using a simple **respirometer**.

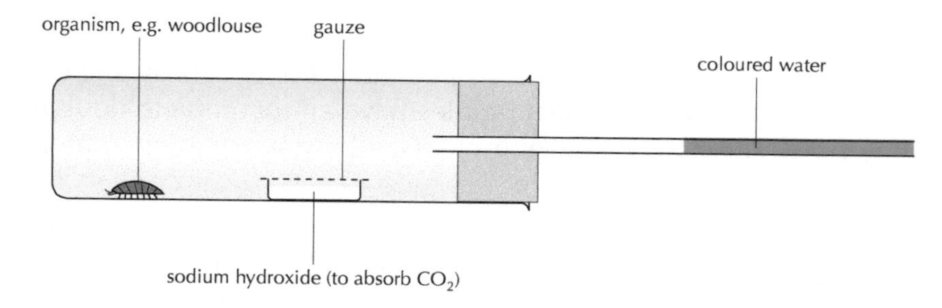

Digital gas probe

TOP TIP

It is possible to use digital gas probes for both carbon dioxide output and oxygen consumption.

Carbon dioxide produced by the living organism is absorbed by the chemical. Oxygen consumed lowers the air pressure in the apparatus and can be quantified by measuring how far along the capillary tube the coloured water has travelled, taking the place of the consumed oxygen.

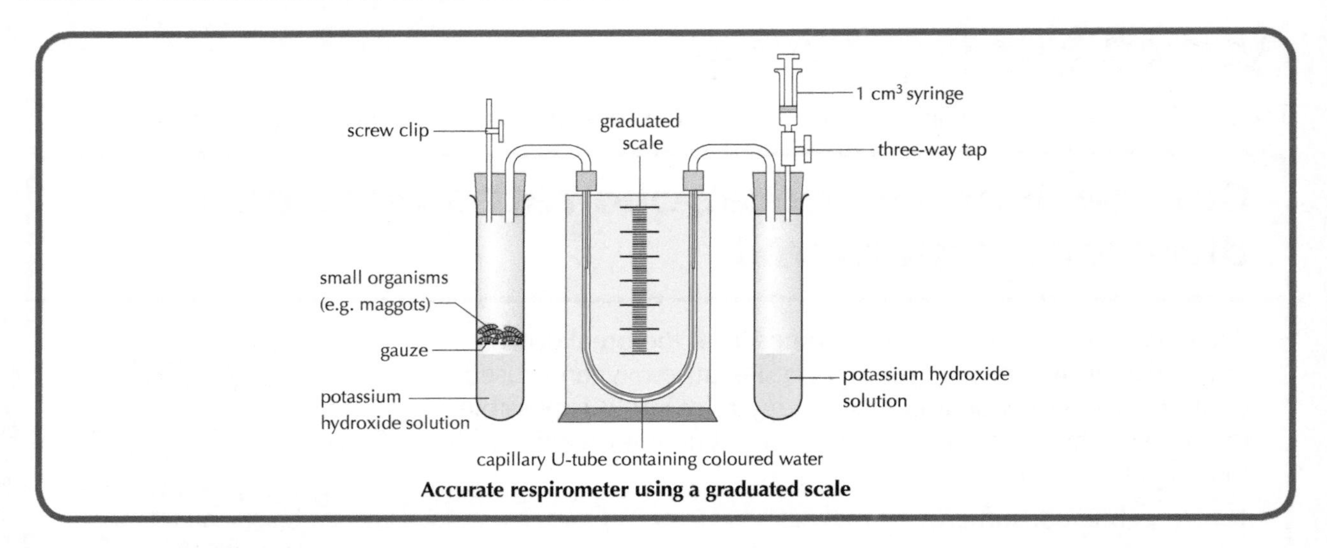

Accurate respirometer using a graduated scale

Efficient oxygen delivery

Complex organisms that have a high metabolic rate have a greater demand for oxygen. These organisms have evolved efficient transport systems to deliver oxygen from the lungs to body cells.

Birds and mammals have a complete double circulatory system consisting of two atria and two ventricles. Amphibians and most reptiles have an incomplete double circulatory system consisting of two atria and one ventricle. Fish have a single circulatory system consisting of one atrium and one ventricle.

Complete double circulatory systems allow delivery of fully oxygenated blood under high pressure to respiring cells. This enables more efficient delivery of oxygen and removal of carbon dioxide and helps higher metabolic rates to be maintained.

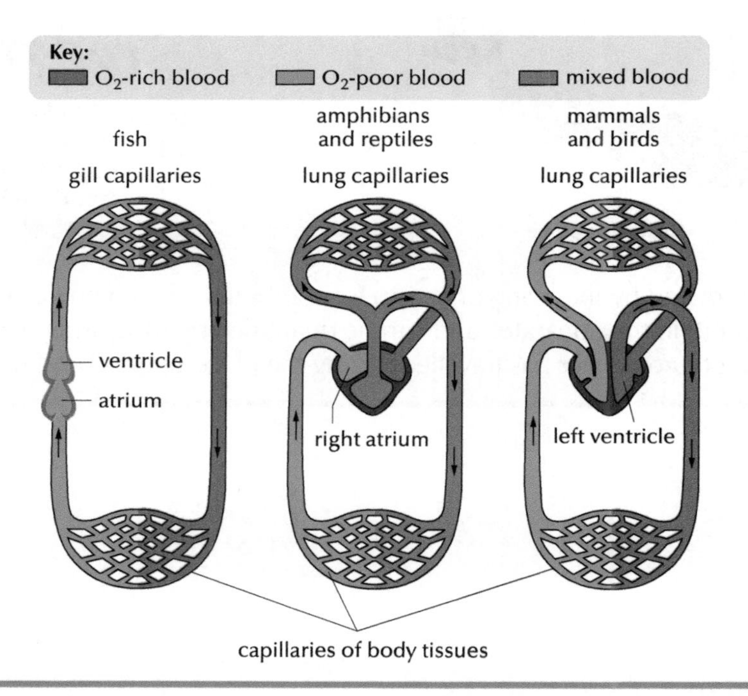

Transport systems of animal groups

Animal group	Number of heart chambers	Type of circulation	Relative rate of metabolism
Fish	Two 1 atrium, 1 ventricle	Single – blood passes through heart once	lowest
Amphibians	Three 2 atria, 1 ventricle	Incomplete double – blood passes through heart twice	Lower (than birds and mammals)
Reptiles	Three 2 atria, 1 ventricle	Double and mostly incomplete	Lower (than birds and mammals)
Birds	Four 2 atria, 2 ventricles	Complete double	highest
Mammals	Four 2 atria, 2 ventricles	Complete double	highest

TOP TIP

In a complete double circulation, blood passes through the heart twice per cycle and there is no mixing of oxygenated and deoxygenated blood. In amphibians and most reptiles, the circulatory system is termed incomplete double. This is because, although blood passes through the heart twice per cycle, mixing of oxygenated and deoxygenated blood occurs in the ventricle – hence 'incomplete'.

Quick Test 18

1. Explain what is meant by metabolic rate.
2. State two variables that can be measured to compare metabolic rates.
3. Explain why a mammalian heart is more efficient at delivering oxygen than a fish heart.
4. State the number of chambers found in both reptiles and amphibians.

Metabolism in conformers and regulators

Effect of external abiotic factors

An organism's metabolic rate is sensitive to abiotic factors in its environment such as:

- temperature
- salinity
- pH.

For an organism to survive, these external abiotic factors need to be within a range that allows the organism's metabolism to function optimally.

Conformers

The **internal environment** of an organism is made up of body cells and their surrounding tissue fluid, which contains water, glucose and other important molecules. This internal environment is very influenced by temperature. The maintenance of the internal environment independently of the external environment is called **homeostasis**.

Conformers cannot regulate their internal environment and therefore cannot control their metabolic rate by physiological means. They are forced to use behavioural responses to maintain an optimum metabolic rate to tolerate variations in their external environment.

> **TOP TIP**
>
> Remember the 3Cs: **c**onformers **c**annot **c**ontrol their internal environment by physiological means.

The body temperature of a conformer depends upon the temperature of the external environment.

Conformers include reptiles and invertebrates.

Conformers have no physiological mechanisms for controlling their internal environment so the metabolic cost is low. However, they can only occupy a narrow range of ecological niches and are less able to adapt to and survive any changes in the external environment.

Regulators

Regulators, such as mammals and birds, use metabolism to control their internal environments. This usually involves physiological mechanisms. The internal environment remains stable despite changes in the external environment.

Regulators can occupy a wider range of ecological niches. Their metabolic costs are high due to the physiological mechanisms needed for homeostasis. The metabolic rate of a regulator is higher than that of a conformer.

Thermoregulation by negative feedback

Mammals are regulators and are able to maintain an internal body temperature of 37°C to facilitate optimum enzyme activity and movement of molecules by diffusion using a system of **negative feedback** control.

- A change in temperature of the internal environment is detected by by **receptors** that generate nerve messages travelling to the **hypothalamus**, the temperature control centre in the brain.
- An electrical impulse sent along nerves communicates information to effectors in the skin.
- The effectors in the skin bring about corrective responses to restore the body temperature to the normal or **set point**.

	Corrective response to increase in body temperature	Corrective response to decrease in body temperature
Receptors	Receptors detect a rise in body temperature and send nerve impulses to the hypothalamus Receptors in the skin monitor an increase in the external temperature	Receptors detect a decrease in body temperature and send nerve impulses to the hypothalamus Receptors in the skin monitor a decrease in the external temperature
Type of signal sent from receptor cells	Electrical impulses along nerves	Electrical impulses along nerves

TOP TIP

Regulators use physiological mechanisms to control their internal environment. These include:

- control of blood glucose
- control of blood water
- control of body temperature.

	Corrective response to increase in body temperature	Corrective response to decrease in body temperature
Response of effectors	• Sweat glands activated so that the water in the sweat evaporates and cools the skin. • Hairs on skin lie flat due to hair erector muscles relaxing • Vasodilation – blood vessels widen, bringing blood to surface of skin increasing the loss of heat by radiation.	• Sweat glands stop producing sweat • Hairs on skin stand up due to contraction of hair erector muscles trapping air as insulation • Vasoconstriction – blood vessels narrow, diverting blood away from surface of skin, thus decreasing heat loss

TOP TIP

Metabolic rate decreases when body temperature increases, and vice versa.

TOP TIP

Remember: body heat is used to evaporate water in sweat, cooling the skin. Remember also that in the cold, shivering causes involuntary contractions of muscles that generate heat.

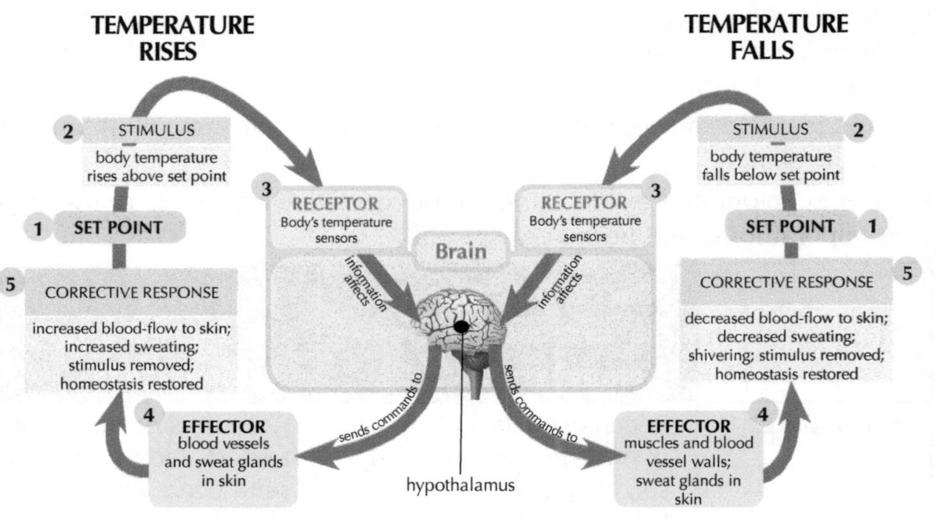

Negative feedback control of body temperature in mammals.

A human is able to regulate body temperature for optimum enzyme function.

Quick Test 19

1. Explain the advantage to conformers of not being able to regulate their internal environment.
2. Explain why conformers can only occupy a very narrow range of environments.
3. State what is meant by a regulator.
4. State one disadvantage of physiological control to regulators.
5. Describe the principle of negative feedback control.
6. Describe the role of the hypothalamus in the control of body temperature.
7. Name one effector that responds to a decrease in temperature.

Many environments change in such a way that normal metabolic activity is not possible. Two main ways in which animals adapt is either to survive adverse conditions or to avoid them completely. Some organisms can survive major changes in environmental conditions such as temperature and drought, through **dormancy** or avoid environmental changes by **migration**.

Dormancy

Dormancy is a stage in the life cycle of an organism where metabolic rate decreases and growth of the organism stops. Entering a dormant state allows the animal to conserve energy when otherwise the energy cost of normal metabolism would not allow it to survive. There are two type of dormancy:

1. **Predictive** – organisms enter into a dormant stage before environmental conditions deteriorate. For example, bears hibernate before the onset of winter.
2. **Consequential** – organisms enter into a dormant stage after a significant detrimental change in environmental conditions.

Lungfish enter a period of aestivation during periods of drought

Particular forms of dormancy include:

Aestivation – a state of inactivity, during periods of high temperature or drought, occurring in some animals such as lungfish, where metabolic processes are slowed down.

Daily torpor – A period of reduced activity in some animals with a high metabolic rate. Such animals can reduce their metabolic rate every 24 hours by becoming inactive and reducing their breathing rate considerably. This helps to conserve energy.

In harsh winters, some animals **hibernate**, which allows them to survive by reducing their heartbeat, body temperature and oxygen uptake. This is coupled with minimal activity and these changes drastically reduce the need for energy by lowering the metabolic rate.

GOT IT?

Migration

Metabolism is sustained in some animals by relocating to another more suitable environment.

This is called migration and allows the animal to avoid adverse conditions.

For example, swallows migrate from Africa to Britain in summer where insects are plentiful, reducing competition for available food. They return to Africa in winter when there are no longer flying insects available as a food source in Britain.

Migrating behaviour in animals is a result of both **innate** and **learned behaviour**.

Behaviour that is innate is genetically based; however, behaviour can also be learned by experience and observing other members of the same species.

Techniques

Scientists use a variety of techniques, such as satellite tracking and leg rings, to study long-distance migration.

Increased use of orbital satellite systems allows an animal to have a transmitter fitted to it that sends signals to a global positioning system.

Ringing has been used for over a century as a way of mechanically tagging birds to monitor their movements over long distances.

Quick Test 20

1. State the type of dormancy an animal enters into following the onset of winter.
2. Give one advantage of daily torpor to small birds and mammals that have a high metabolic rate.
3. State the type of behaviour shown by animals to avoid metabolic adversity.
4. Explain the difference between innate and learned behaviour.
5. State one advantage of using satellite tracking compared with ringing.

Environmental control of metabolism

Growing micro-organisms

Metabolism of micro-organisms, which are archaea, bacteria and some species of eukaryotes, can be controlled during culture in the laboratory by changing environmental conditions in order to increase the final level of product.

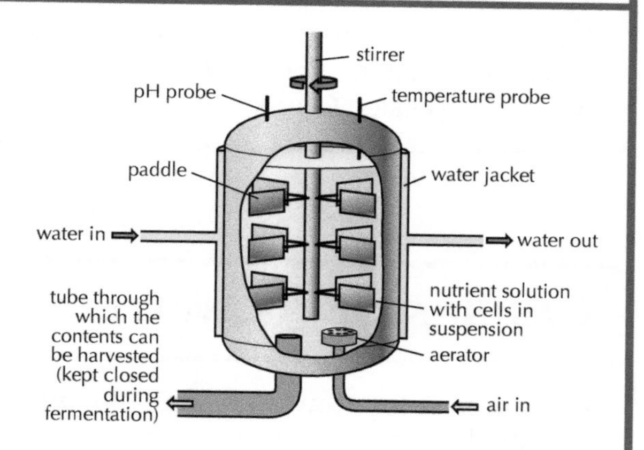

A growth chamber, called a **fermenter** may be used. This chamber allows the medium in which the micro-organism is growing to be closely regulated, often by computer control in terms of pH, nutrient levels etc. The sterile conditions in the fermenter reduce competition for nutrients by other micro-organisms.

Culturing conditions that can be altered are:

1. glucose concentration
2. temperature
3. pH levels using buffer solutions
4. oxygen concentration by aeration of growth medium.

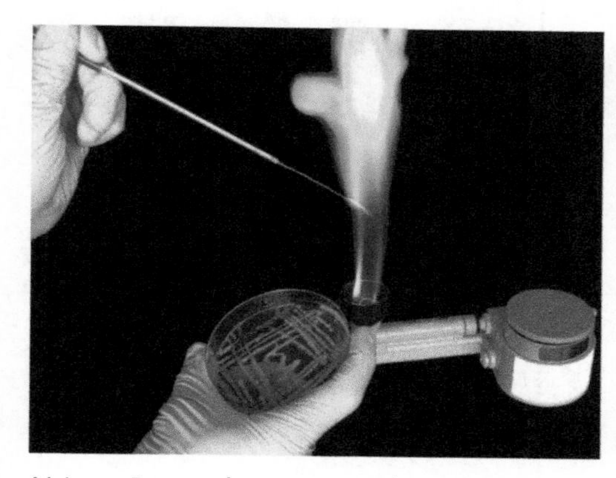

Using a Bunsen burner to sterilise

Micro-organisms must be grown in **aseptic** conditions, to eliminate the growth of wild micro-organisms in the air, which may contaminate the culture medium.

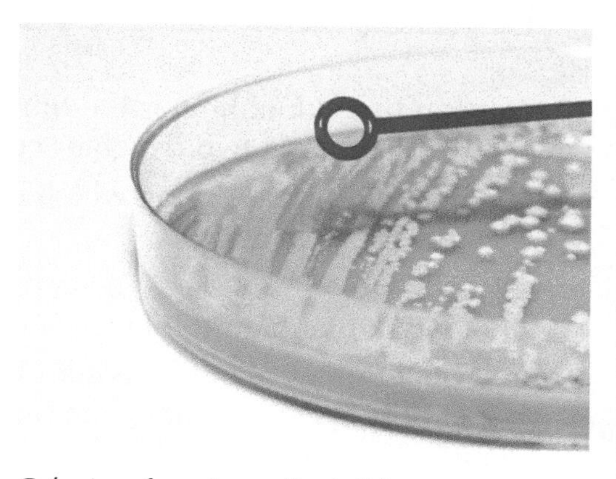

Colonies of yeast on a Petri dish

Micro-organisms produce all the complex molecules required for **biosynthesis**. These include amino acids, vitamins, fatty acids, etc. Other micro-organisms require these to be put into the growth medium.

TOP TIP

Photosynthetic micro-organisms derive their energy from light.

Usefulness of micro-organisms

The specific metabolism of micro-organisms may allow them to be used industrially or in geochemical processes. For example, some bacteria use methane gas as a respiratory substrate and so where its production from rice paddy fields, refuse dumps and swamps, etc could cause air pollution, these bacteria can detoxify the methane gas.

Micro-organisms are widely exploited by humans because they grow so quickly, are adaptable and easy to culture. Recent developments in genetic engineering have seen some bacteria being manipulated to produce antiviral drugs, cancer treatments and insecticides, etc.

Phases of growth in cell cultures

If conditions for growth are suitable, micro-organisms can grow very rapidly. Bacteria can double their numbers every 20 minutes.

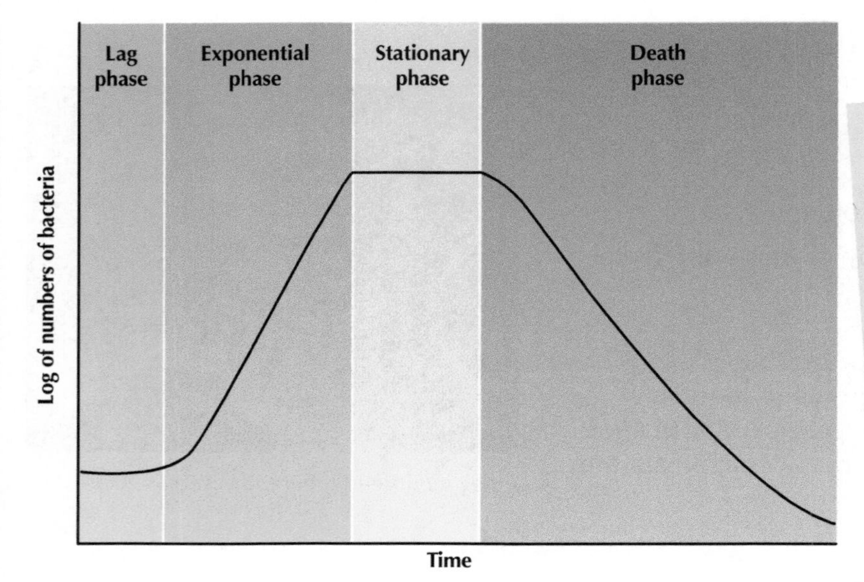

TOP TIP

When a sample of a cell culture is examined under a microscope:
- the **total cell count** involves counting viable and non-viable cells
- the **viable cell count** involves counting only the living micro-organisms.

Generation time (doubling) is the time taken for one cell to divide by mitosis into two daughter cells.

- **Lag phase** – period during which enzymes are induced to metabolise substrate.
- **Exponential (log) phase** – the cells divide at the maximum rate due to plentiful nutrients.
- **Stationary phase** – nutrients start to get used up and toxic end-products accumulate. The number of micro-organisms being produced is balanced by the numbers dying.
- **Death phase** – eventually, the micro-organisms are starved of nutrients or cannot tolerate their own toxic wastes, so they die.

TOP TIP

In the stationary phase, secondary metabolites such as antibiotics are produced. In the wild, these confer an ecological advantage by allowing the micro-organism that produced them to out-compete other micro-organisms.

Growth curves of micro-organisms

The growth of micro-organisms can be plotted in different ways. A viable count can show the lag and stationary phases but the numbers increase so rapidly, it becomes impossible to continue the plot.

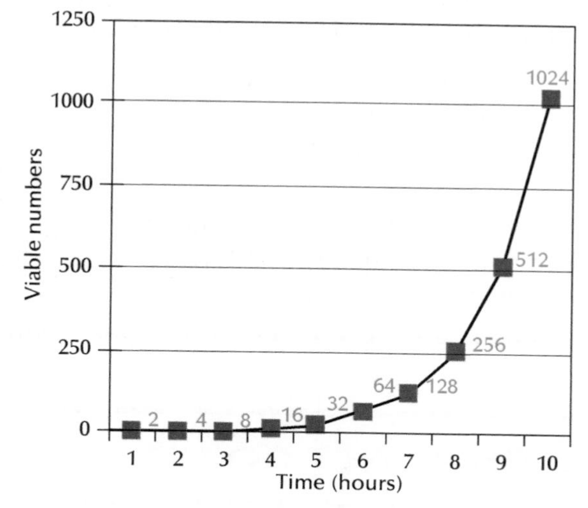

Using a different scale on the y-axis and special semi-logarithmic graph paper, it is possible to plot accurately the increasing numbers over several hours.

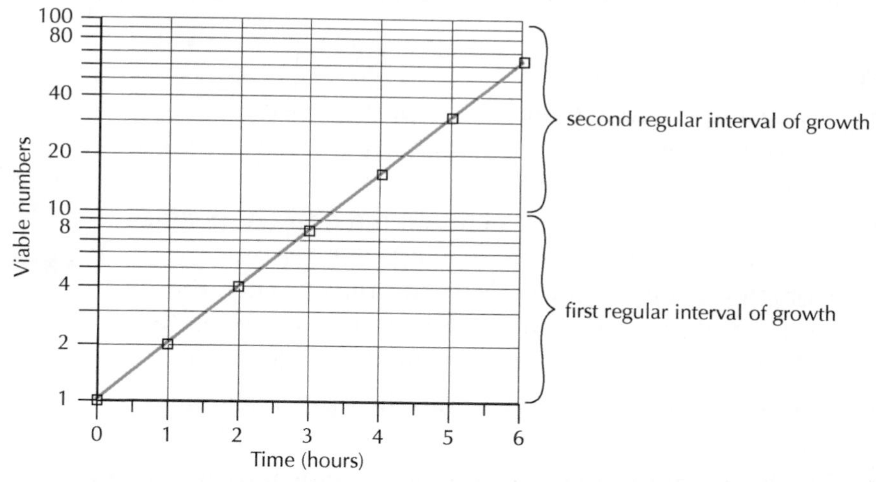

Quick Test 21

1. State three variables that can be altered when growing micro-organisms to maximise growth.
2. Identify two molecules that must be present in the growth medium to allow biosynthesis to take place.
3. Explain what is meant by the 'generation time' in a cell culture.
4. State what happens to a cell culture during the 'exponential phase'.
5. Explain why the numbers of micro-organisms remain constant during the 'stationary phase' in the growth of a cell culture.

Genetic control of metabolism

Improving wild strains of micro-organisms

Many wild strains of micro-organisms have desirable phenotypes typically by expressing a product that is useful. It is now possible to improve these wild strains by:

1. **mutagenesis** – using agents such as ultra-violet radiation or mutagenic chemicals to induce a mutation resulting in the production of an improved strain of micro-organism
2. **recombinant DNA technology** – allows genetic material to be transferred between different organisms, allowing the micro-organism to code for a useful protein.

TOP TIP

Wild strains of micro-organisms are often exploited in both industry and research because they grow so rapidly, are adaptable and are easy to culture.

Recombinant DNA technology

Recombinant DNA technology is the process of transferring gene(s) from one species to another. This involves the use of plasmids and **artificially produced chromosomes** as **vectors**.

Process of gene transfer.

- A plasmid in a bacterial cell is cut open using special enzymes called **restriction endonucleases** that cut at a specific base sequence called a **restriction site**.
- Restriction endonucleases cut out the desired specific gene from a chromosome, leaving **sticky ends**. Complementary sticky ends are produced when the same restriction endonuclease is used to cut open the plasmid and extract the gene from the chromosome.
- The desired gene is inserted into the open bacterial plasmid and sealed in place by the enzyme ligase.
- The bacterial plasmid containing the new gene is called a recombinant plasmid.

TOP TIP

Artificial chromosomes are preferred to plasmids as vectors because they can carry a lot more genetic material.

TOP TIP

Remember:

- DNA restriction endonuclease acts as 'scissors', cutting out a gene.
- Ligase acts as 'glue', sealing the transferred gene into place.

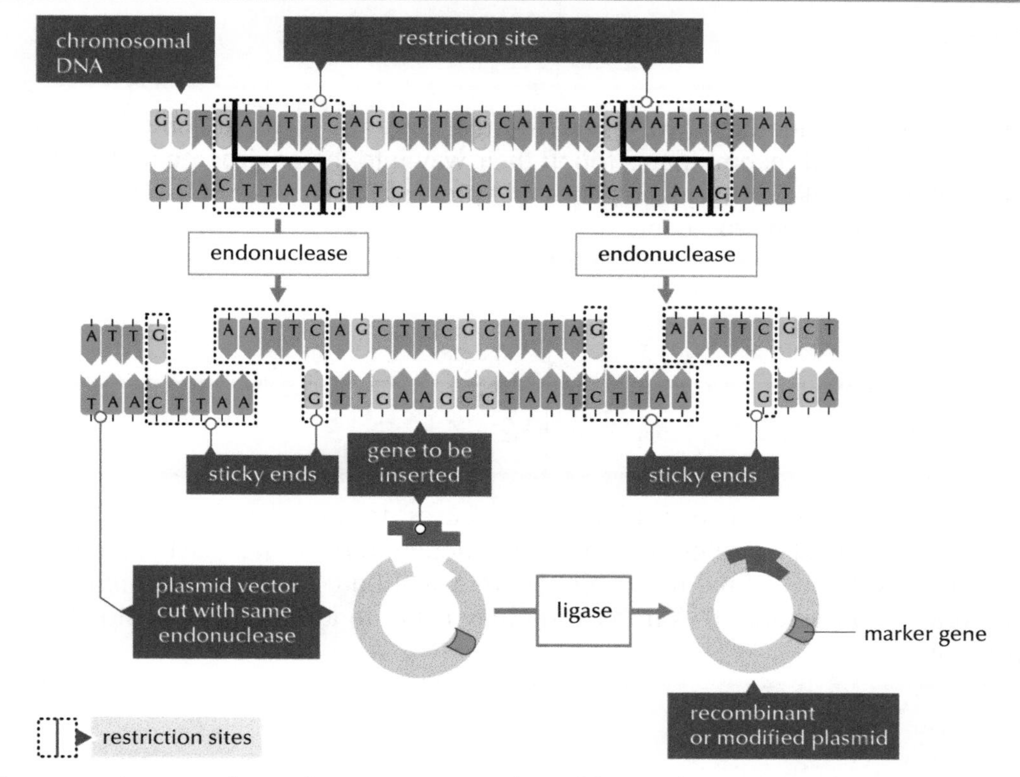

Usually, recombinant plasmids carry some **selectable markers** that allow the modified host cells that have taken up those plasmids to be identified.

Further, the plasmid will have incorporated an **origin of replication**, which consists of genes able to control self-replication of the plasmid DNA. Also present are **regulatory gene sequences**.

TOP TIP

Two common selectable marker genes alter the host cell, making it resistant to antibiotics or making it glow under special lights if it has incorporated the recombinant plasmid.
Regulatory gene sequences can speed up gene transcription, increasing the rate of protein synthesis.

TOP TIP

Recombinant plasmids and artificial chromosomes both contain restriction sites, regulatory sequences, an origin of replication and selectable markers.

TOP TIP

Selectable marker genes, such as those conferring antibiotic resistance, protect the micro-organism from a selective agent such as an antibiotic that would normally kill it or prevent it growing.

Safety

Recombinant gene technology involves micro-organisms and so there could be concern of the potential effect if recombinants found their way into the environment. As a safety precaution, it is possible to introduce genes that prevent the survival of these recombinant micro-organisms in the environment.

Use of recombinant yeast cells

Instead of using bacteria, it is possible, in some cases, to use recombinant yeast cells to produce active forms of a protein that may be inactive in bacteria. Yeast is a eukaryote and therefore plant and animal recombinant DNA will be correctly folded in recombinant yeast whereas recombinant DNA in bacteria may result in an incorrectly folded eukaryotic protein.

Quick Test 22

1. Give one example of how a wild strain of a micro-organism can be improved.
2. State which part of a bacterial cell's genetic material is frequently used in recombinant DNA technology.
3. Explain the role of a restriction endonuclease enzyme in the process of gene transfer.
4. Explain the function of a marker gene.

Food supply, plant growth and productivity

Food supply

Food supply is the provision and distribution of food to a consumer.

Food security is the ability of human populations to access food of sufficient quality and quantity.

Food production must be sustainable so that people have access to quality food over long-term timescales. Production of food must not degrade the natural resources on which agriculture depends.

TOP TIP

As the global population increases, demand for food will also increase.

Food availability depends upon efficient, sustainable agricultural methods of producing food crops.

Crop production depends upon the ability of soil to support plant growth, together with environmental factors that affect photosynthesis such as temperature and water availability.

In developing countries, where there is poor soil and long periods of drought, people have very little food security.

Producing crops in a limited area

Production of food crops such as wheat, maize, potatoes, roots, legumes and rice grown on a small area of land can be increased by:

- growing a genetically high yielding species of crop plant or **cultivar**
- applying nitrogen-based fertilisers
- reducing pests and diseases that attack the crop
- reducing competition between growing crop plants.

Plant growth and productivity

Food production depends on factors that affect photosynthesis and therefore the growth of plants. An increase, for example, in abiotic factors such as light intensity, temperature and carbon dioxide concentration will cause an increase in the rate of photosynthesis. The use of chemical fertilisers and pesticides, breeding of higher-yielding cultivars, competition and the effect of disease as well as the area of land available to grow crops will all impact on food production.

TOP TIP

Productivity is affected by:

- light intensity
- abiotic factors such as temperature, water availability
- soil type
- efficiency of leaves at capturing available light – leaf shape, angle, area
- crop density
- availability of nutrients.

Livestock productivity

Livestock are inefficient converters of the energy contained in plants to animal tissue and so produce less food per unit area. This is due to the loss of energy between trophic levels. However, livestock production is often possible in habitats that are not suitable for growing crops.

TOP TIP

To help support sustainable food production, breeders seek to develop crops with:

- higher nutritional value
- resistance to pests and diseases
- physical characteristics such as length of stalk, for appropriate harvesting
- physical characteristics that can help crops thrive in particular environmental conditions.

Quick Test 23

1. State what is meant by 'food security'.
2. Suggest two ways in which the yield of a crop, grown on a small area of land, may be increased.
3. State two abiotic factors that affect plant productivity.
4. Explain why eating plants directly as food is more efficient that eating animals as food.

Photosynthesis

Light energy from the sun can be absorbed, reflected or transmitted. Some of the absorbed light is used in photosynthesis to generate glucose and oxygen.

Photosynthetic pigments

Plants contain coloured chemical compounds called **pigments**. These photosynthetic pigments are found in stacked membranes within the chloroplast. Each pigment absorbs different wavelengths of light, which make up the visible spectrum. These pigments are found on the inner membranes of chloroplasts where the light-dependent stage of photosynthesis, called photolysis, takes place. ATP is made during this process from ADP and P_i.

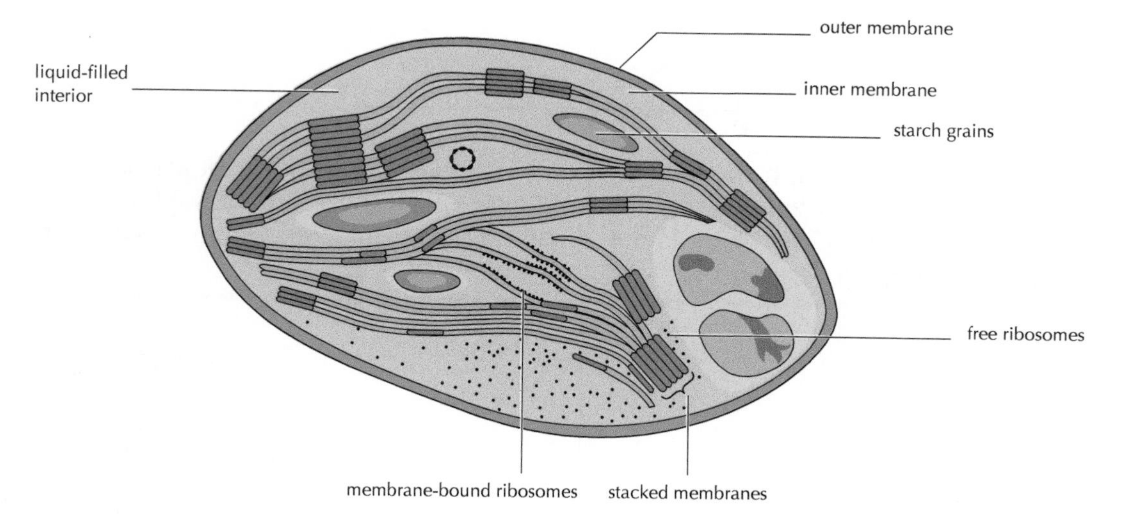

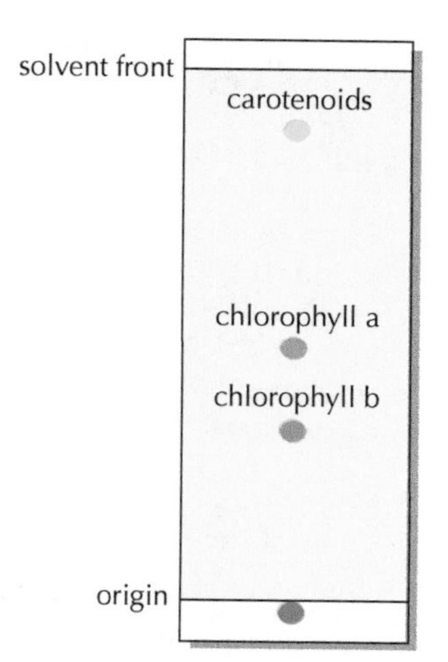

Photosynthetic pigments in a green plant separated by chromatography

TOP TIP

Chlorophyll a and b absorb most light energy in land plants.
Carotenoids are accessory pigments that pass on absorbed light energy to chlorophyll a.

The main photosynthetic pigments are:

1. chlorophyll a
2. chlorophyll b
3. **carotenoids**

Pigments can be extracted from a green plant and identified using the process of thin layer chromatography to produce a chromatogram. The pigments have different solubilities in the solvent. Those that are most soluble travel furthest, those that are least soluble are found nearer the origin of the chromatogram.

Chlorophyll a and b absorb most light energy at the blue and red end of the spectrum.

The carotenoids extend the absorption spectrum by absorbing light of different wavelengths compared with chlorophyll a and b.

Absorption spectrum

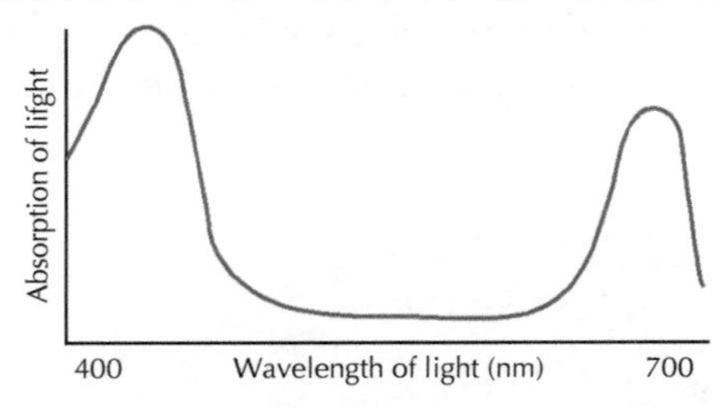

An **absorption spectrum** graph shows which colours (wavelengths) of light are absorbed by each of the three photosynthetic pigments. Plants absorb most light at the blue and red end of the spectrum. Green light is not absorbed but is reflected which is why many plants appear to be green.

Action spectrum

An **action spectrum** graph charts the effectiveness of different wavelengths at 'driving' photosynthesis. The rate of photosynthesis is highest at the blue and red ends of the spectrum where most light is absorbed by green plants. There is a little photosynthetic activity in the yellow and orange parts of the spectrum due to some absorption of these wavelengths of light by the carotenoid pigments.

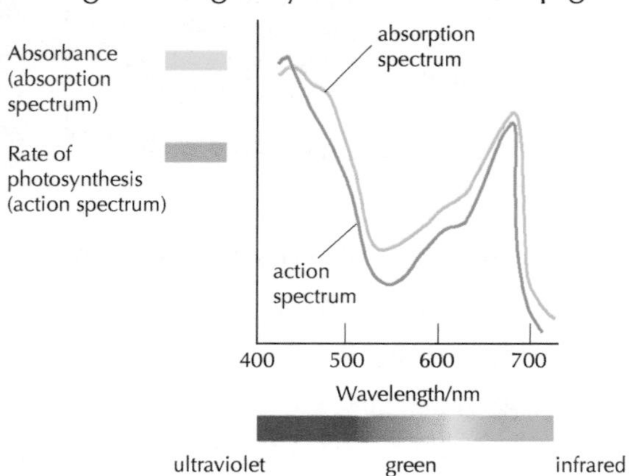

TOP TIP

Remember that the 'action' referred to in the action spectrum is photosynthesis!

TOP TIP

Not all light landing on a leaf is absorbed into the cells:

- some light passes straight through the leaf – transmitted
- some light bounces off the surface of the leaf – reflected.

TOP TIP

Plants that grow in shady areas underneath the leaf canopy in forests rely on transmitted and reflected light to survive.

Process of photosynthesis

The manufacture of glucose in plant cells using the raw materials of water and carbon dioxide gas together with light energy occurs in two stages. The first stage, photolysis, takes place on the stacked membranes of the chloroplast. The second stage, the **Calvin cycle**, takes place in the liquid interior of the chloroplast.

Photosynthesis involves the following processes:

- Absorbed light energy excites electrons within pigment molecules.
- Some excited electrons transfer through the electron transport chain, rotating the enzyme molecule ATP synthase, located in the membranes of the chloroplast generating a molecule of ATP from ADP + P_i.
- Other excited electrons break the chemical bonds between hydrogen and oxygen atoms in a molecule of water.
- Oxygen is evolved and hydrogen is released.
- Oxygen moves though the plant cells by diffusion and exits the leaf through the stomata.
- Hydrogen is picked up by the co-enzyme NADP, to form NADPH.
- Both NADPH and ATP produced during photolysis are essential to the next stage of photosynthesis, called carbon fixation.

> **TOP TIP**
>
> Oxygen is used for AEROBIC respiration. It is also used for respiration in plant cells.

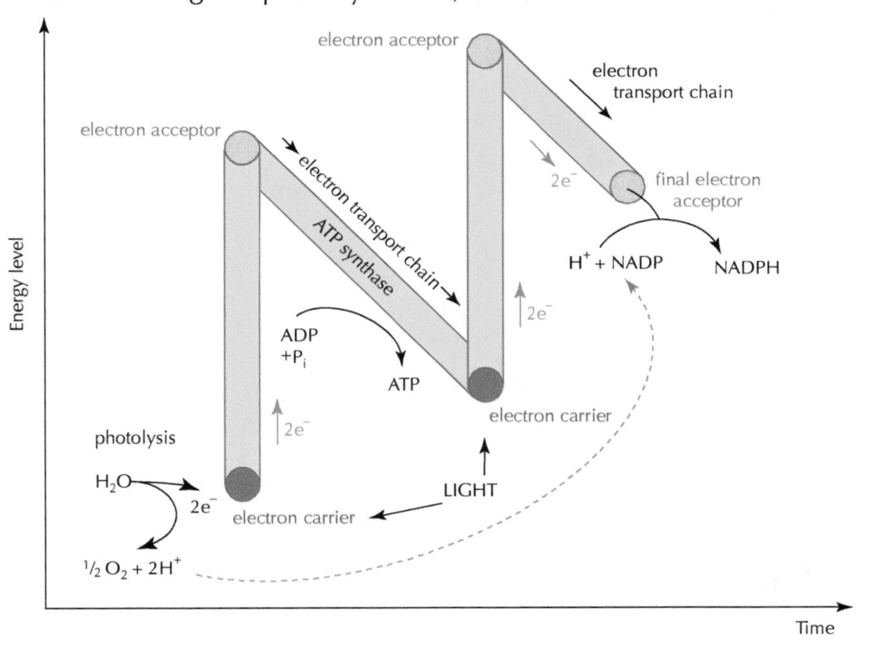

Carbon fixation:

- is light independent
- occurs in the liquid-filled interior of the chloroplast, and is dependent upon NADPH and ATP being produced during photolysis.

Stages of carbon fixation:

- Carbon dioxide enters the leaf through the stomata, and moves by diffusion into the chloroplast.

- Carbon dioxide combines with **RuBP** (ribulose biphosphate) catalysed by the enzyme **RuBisCo** (ribulose biphosphate carboxylase oxygenase).
- 3-phosphoglycerate (3PG) is phosphorylated and reduced and, using the ATP and NADH from photolysis, is converted into glyceraldehyde-3-phosphate (G3P).
- G3P has two possible fates; some is used to regenerate RuBP, which is continually being used up, and some is used to make glucose.
- Glucose can then be used in cell respiration, stored as starch grains in the chloroplast or polymerised to make cellulose for new cell walls.

TOP TIP

The co-enzyme that picks up hydrogen in cell respiration is NAD, and in photosynthesis it is NADP – remember P for photosynthesis!

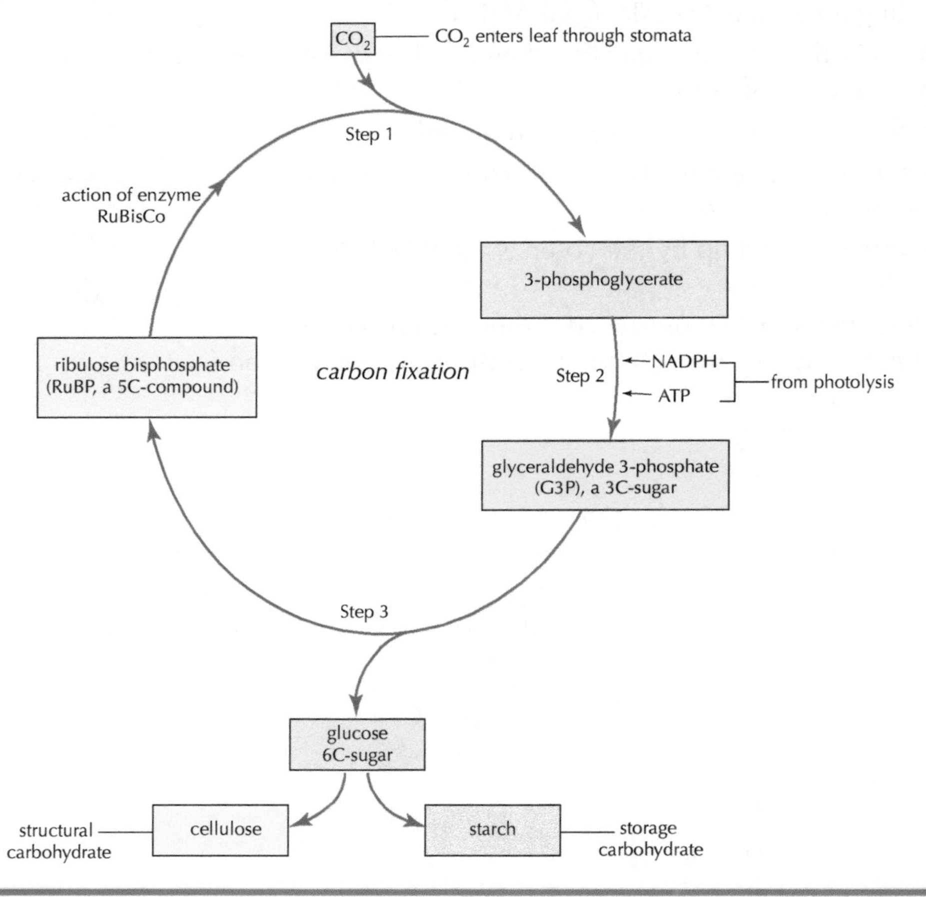

Quick Test 24

1. State the pigments that absorb most of the light used in photosynthesis.
2. Explain the difference between an absorption and action spectrum.
3. State the two molecules, produced during photolysis, that are essential for the Calvin cycle.
4. Name the enzyme that catalyses the reaction between carbon dioxide gas and RuBP.
5. State the two possible fates of glyceride-3-phosphate produced in the Calvin cycle.
6. Name the hydrogen carrier in photosynthesis.

Plant and animal breeding

Improving productivity

Productivity of plants and animals in agriculture in order to provide a sustainable source of food can be improved through genetics. By breeding only those plants and animals that have desirable phenotypes within each subsequent generation, an improved breed can be sustained.

TOP TIP

Improvements in livestock include resistance to pests and diseases, higher nutrional values and yields and the ability to thrive in particular environments.
Improvements in crops include resistance to pests and diseases, higher nutritional values and yields and ability to thrive in particular environmental conditions.

Field trials

Different species of crops, such as grass, cereals and potatoes, can be grown on identical small areas of land, called plots. Each set of plots are exposed to different treatments, then harvested to estimate the yield. This information allows farmers to select a species that will give the highest yield within local growing conditions.

Field trials can also be used to generate data on the performance of **genetically modified crops**, or new cultivars. Such trials can be used to gather data on the effects of treatments, such as different levels of fertiliser application on the growth of crops. Comparative data from different plots can then be analysed.

TOP TIP

Field trials have three main components that need to be taken into account:

1. selection of treatments – for valid comparisons
2. randomisation of treatments – to eliminate bias
3. number of replicates – to account for the variability within the sample.

TOP TIP

Possible desirable genetic chacteristics:

- plants – insertion of Bt toxin gene results in plants with increased pest resistance, insertion of glyphosate resistance gene into plants has resulted in herbicide tolerance
- animals – high conversion of food to body mass, high fertility rate, disease resistance.

Inbreeding

In farming systems, animals and plants can be inbred through continued selective breeding programmes. **Inbreeding** involves crossing closely related individuals until the population breeds true to the desired phenotype due to the elimination of heterozygotes.

Breeding genetically similar plants or animals can result in an increase in the frequency of harmful homozygous recessive alleles in the genotype. These individuals do less well at surviving to reproduce. This is called **inbreeding depression** and is a consequence of heterozygosity being reduced.

TOP TIP

Inbreeding depression can be avoided by introducing new individuals to maintain genetic diversity.

Cross breeding and F_1 hybrids

Crossing individuals of different breeds within a species can result in both genetically and physically stronger F_1 offspring, and can produce desirable characteristics from each parent. This is called **hybrid vigour**.

Scottish Half Bred sheep are a cross between Cheviot and Border Leicester breeds. The Half Bred has a large body frame inherited from the Border Leicester and high muscle mass inherited from the Cheviot.

Cheviot

Border Leicester

In plants, F_1 hybrids produced by the crossing of two different inbred lines create a relatively uniform heterozygous crop with increased vigour, yield, disease resistance and growth rate.

F_1 hybrid cherry tomatoes with many fruits and fungal disease resistance

TOP TIP

In inbreeding animals and plants, F_1 are not usually bred together as the F_2 produced shows too much variation.

TOP TIP

In animals, more cross breeds showing the improved characteristic can be produced by maintaining the two parent breeds.

Genetic technology

By identifying the base sequences of an organism's genome using PCR and bioinformatics, desirable gene sequences can be identified, directing the management of breeding programmes. If desirable gene sequences are located, the organism will be selected and used within future breeding programmes. This ensures the improved characteristics are passed on to subsquent generations of animals and plants.

TOP TIP

Using recombinant DNA technology, individual genes can be inserted into the genomes of crop plants resulting in new genetically modified plants that will grow with the improved characteristic.

Genome sequencing

Using PCR and bioinformatics, the entire genome of an organism can be analysed. Base sequencing can then be carried out to produce a genome sequence. This sequence allows the presence or absence of a desirable allele of a gene to be determined.

Quick Test 25

1. Explain how field trials provide a useful way of selecting desirable genetic characteristics in plants.
2. Explain how inbreeding depression can occur.
3. Explain how crossbreeding animals of different breed within the same species can be advantageous.
4. Explain why using F_1 hybrids is not commonly used in breeding programmes to produce an F_2 generation.
5. Describe how genomic sequencing can be used to improve animal breeding.

Crop protection

Weeds, pests and diseases

Productivity of a crop can be greatly reduced by:

- weeds
- pests and diseases.

Weeds

Annual weeds	Perennial weeds
• Short life cycle (one year). • Rapid growth. • High numbers of seeds produced. • Dormant seeds remain viable for long period of time. • Examples: goose grass, knotweed, shepherd's purse.	• Long life cycle (two years+). • Capable of **vegetative reproduction**. • Storage organs provide food for plant in autumn/winter. • Examples: couch grass, nettles, buttercup.

TOP TIP

A 'weed' is a plant growing in the wrong place!

TOP TIP

Weeds within a growing crop can:

- compete with crop plants for nutrients
- contaminate crop at harvest time
- release chemicals into soil that inhibit growth of crop plant
- provide habitats for pest species.

Pests and diseases

Pests	Diseases
• Insects – cause leaf damage, such as greenfly, blackfly, leatherjackets. • Nematodes – round worms in soil attack roots and storage organs. • Molluscs – slugs and snails damage leaves.	• Fungi – yellow rust on leaves of cereal plants; brown rot on stone fruits, peaches and plums. • Bacteria – affect stems, roots and leaves and cause leaf spots, blight and galls. • Viruses – yellow mosaic of lettuce leaves caused by the lettuce mosaic virus reduces photosynthesis.

TOP TIP

Plant diseases, caused by fungi, bacteria or viruses are often transmitted by invertebrates.

Cultural control methods

Cultivation and good management of soil help to prevent the build up of weeds, pests and diseases.

1. Ploughing – this destroys perennial root systems in the soil and cleans the field of any previous crop residue that may harbour some pests and plant diseases.
2. Weeding – growing weeds are removed by cultivation between rows of crop plants.
3. Crop rotation – planting different crops in each field each year helps to break the life cycle of pest species.

TOP TIP

Pesticides include insecticides that kill insect pests, molluscicides that kill snails and slugs, and nematicides that kill the small nematode worms.

Chemical control methods

The action of a pesticide can be selective or systemic.

Systemic Pesticide	Selective pesticide
Spreads throughout the animal body or the entire vascular system of a plant.	Attacks only particular target pests.

TOP TIP

Application based on disease forecasts is more effective than treating a crop that already has the disease.

TOP TIP

Pesticides can be designed to be highly specific as to the target pest while degrading quickly and thus avoiding persistence.

Disadvantages of chemical crop protection:

- Toxic to non-target species such as wild animals.
- Chemical remains in soil and surrounding environment following harvest of crop.
- Chemical enters food chain and accumulates in the body tissue of animals. The concentration of the chemical increases through the trophic levels of a food chain, with the predator containing the highest concentration of chemical in body tissues such as fat. This is called **biomagnification**.
- Repeated exposure to the same chemical can select for resistant strains of pests and diseases.

Biomagnification of pesticide DDT in a food chaim

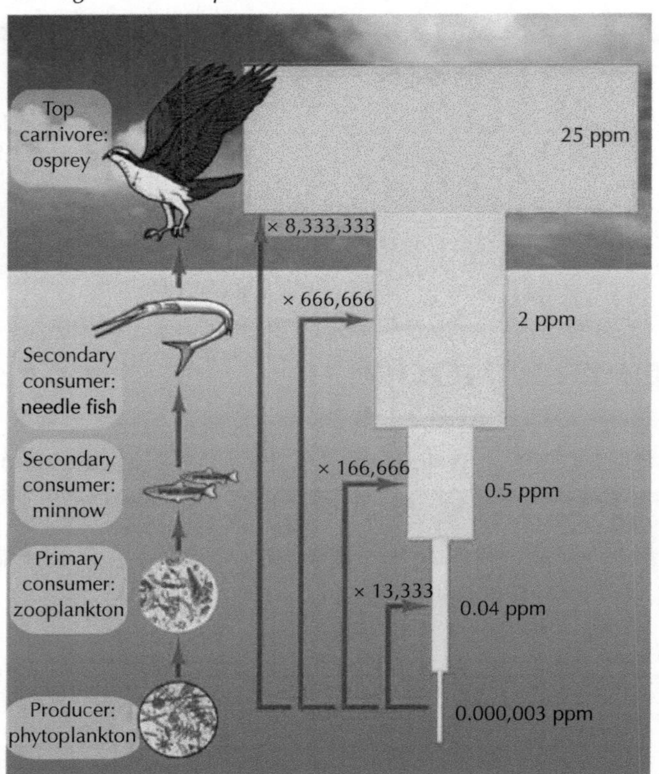

TOP TIP

Bioaccumulation occurs when there is a build-up of a chemical in an organism.

Biological control and integrated pest management

The use of an integrated pest management scheme reduces:

- the need for chemicals
- the pest population by the deliberate introduction of a natural predator, parasite or pathogen of the pest. For example: a greenfly infestation in a glass house can be biologically controlled by introducing ladybirds that eat greenfly.

Risks of biological control:

- Predator, parasite or pathogen used to control a pest population could also reduce populations of native species, causing ecosystems to collapse. A predator may feed on more species of prey than the target species, causing imbalance in a food web. The control organism itself can become an invasive species, a parasite of other species or prey on or be itself a disease-causing agent.

Ladybird attacking aphids on an endangered plant

TOP TIP

Integrated pest management (IPM) uses a combination of chemical, cultural and biological techniques to control the population size of a pest, instead of killing the whole population.

In Australia, an invasive species of moth has been introduced that feeds on the prickly pear cactus. This controls the spread of the prickly pear. Another example of an invasive species in Australia is the cane toad.

Quick Test 26

1. State two factors that may compete with a growing crop, reducing productivity.
2. State three differences between annual and perennial weeds.
3. State two examples of invertebrate crop pests.
4. Describe how cultivation techniques in agriculture may be used to control weeds, pests and diseases in crops.
5. Identify two main groups of weed killer.
6. State two disadvantages of biological pest control.

Animal welfare

Costs, benefits and ethics

The care and welfare of animals farmed for food is not only ethically desirable, but has been shown to increase productivity.

Cost	Benefit
Financial investment to improve environmental living conditions for animals.	Animals are less stressed, leading to greater productivity and reproductive rate.

TOP TIP

Intensive farming systems reduce costs in labour and feed but evidence shows the stress caused to the animals causes reproductive failure. It is generally deemed to be less ethical with poor animal welfare but generates higher profits than free range farming.

Indicators of animal stress

Tiger in a cage

- Stereotypy – repetitive and unusual behaviour patterns directed towards the environment are observed; for example, animal pacing to and fro in a cage.
- Misdirected behaviour – a normal behaviour, such as preening feathers in birds, is directed against the animal itself, leading to excessive preening and pulling out of feathers.
- Failed sexual/parental behaviour – animal stress as a result of isolation or living in a poor environment can inhibit reproductive behaviour; young produced under these conditions may be rejected by the parent.
- Altered levels of activity – animals suffering from stress may be recognised by excessive high or low levels of activity in the form of hyper-aggression or excessive sleeping.

Quick Test 27

1. State two benefits of investing money into environmental improvements for farm animals.
2. (a) State four main indicators of animal stress.
   (b) State which of these indicators results in an otherwise normal behaviour being used by the animal against itself and causing harm.

Parasitism

In a parasitic relationship, one partner benefits while the other is harmed. The parasite benefits from nutrients and energy provided by the host animal, which is harmed by the loss of these resources.

Parasites often have limited metabolism and cannot survive out of contact with a host.

Head of a human tapeworm

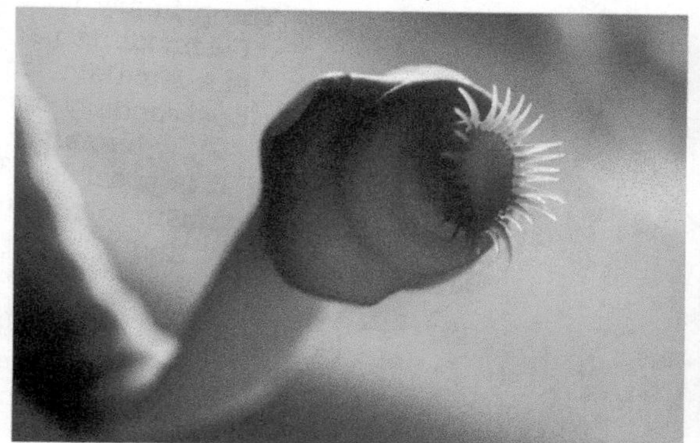

Tapeworm parasite

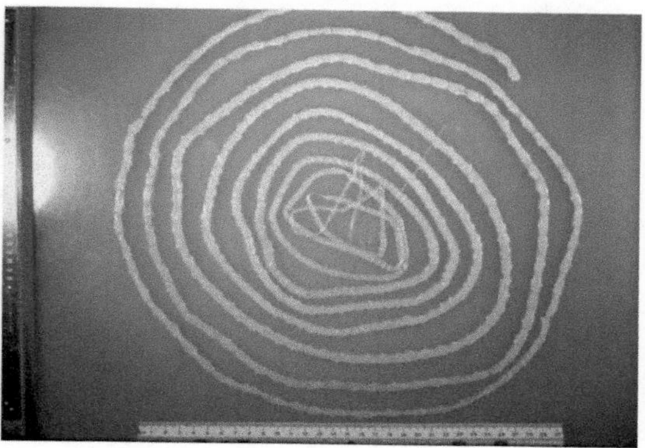

Transmission of parasites

- **Direct contact** – host animals have physical contact with each other and pass the parasite between them. For example, the human head louse is transmitted by direct head-to-head contact.
- **Resistant stage** – for example, cat flea larvae are a resistant stage of the parasite that can be picked up from the environment by a new cat host.
- **Vector** – this refers to an organism that transports the parasite from one host to another. For example, the mosquito acts as an invertebrate **vector organism** for the malaria parasite.

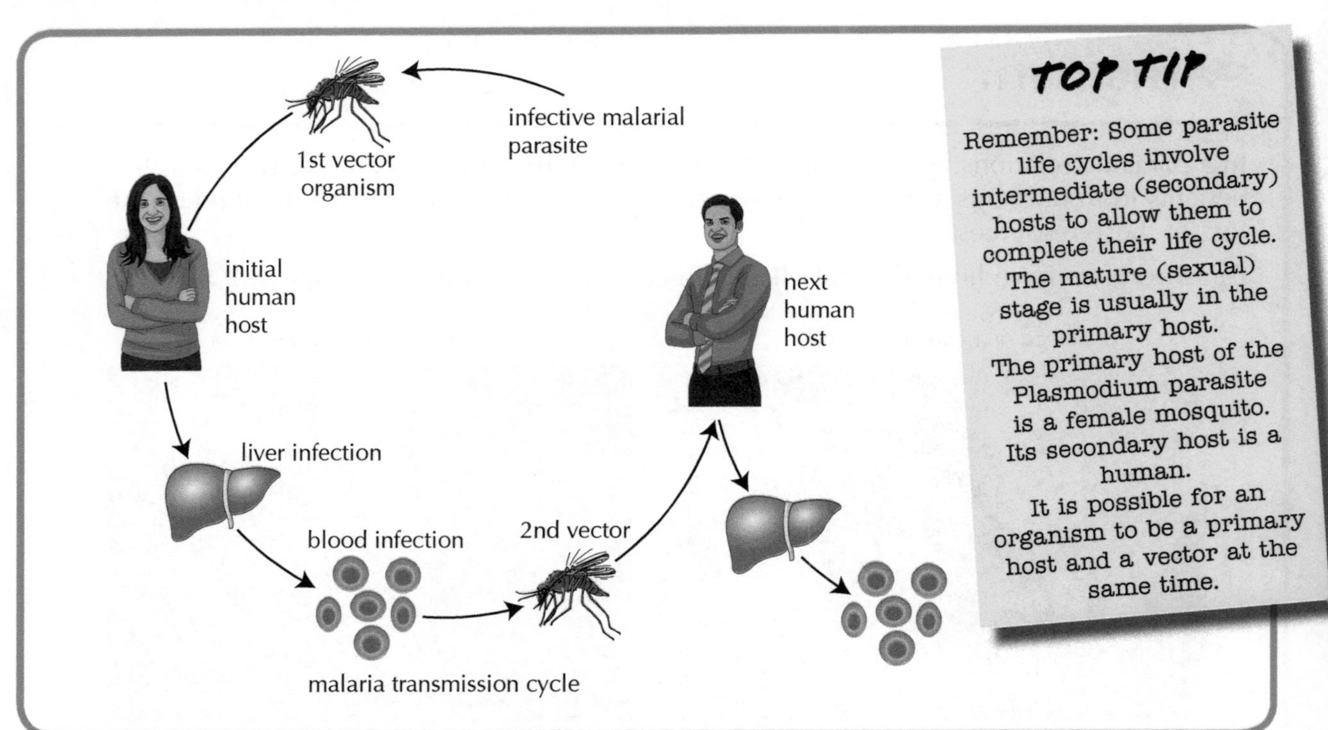

TOP TIP

Remember: Some parasite life cycles involve intermediate (secondary) hosts to allow them to complete their life cycle. The mature (sexual) stage is usually in the primary host. The primary host of the Plasmodium parasite is a female mosquito. Its secondary host is a human. It is possible for an organism to be a primary host and a vector at the same time.

Mutualism

Mutualism is a type of symbiotic interdependent relationship where both partners benefit.

Lichen

TOP TIP

A lichen is an example of a mutualistic relationship between algae and a fungus which grow together in one unified structure. Algae receive nutrients from the fungus, and the fungus in turn receives glucose from the photosynthetic algae.

The oxpecker bird feeds on small insects in the coat of a buffalo. The buffalo benefits from having pests removed, while the oxpecker bird benefits from a plentiful source of food.

Oxpecker birds and buffalo

Clownfish live among the stinging tentacles of sea anemones. A thick mucus layer on the skin protects the fish from stings, which deter predators. The sea anemone is protected as the presence of the clownfish deters the butterfly fish which eat sea anemones. Both partners benefit.

Clown fish and sea anemone

Social behaviour

Many animals live in social groups and have behaviours that are adapted to living in a group such as:

1. social hierarchy
2. co-operative hunting
3. social defence.

Social hierarchy

Social hierarchy is a rank order within a group of animals consisting of dominant and subordinate members. Positions range from the strongest to the weakest individual within the group. Social hierarchy can be seen in packs of wolves and social groups of chimpanzees. In a social hierarchy, dominant individuals carry out ritualistic (threat) displays whilst subordinate animals carry out appeasement behaviour to reduce conflict.

Dominant and submissive wolves interacting

Advantages of social hierarchy

Social hierarchies increase the chances of a dominant animal's favourable genes being passed onto offspring. The formation of alliances within hierarchies helps individuals increase their social status within the group.

Cooperative hunting

Animals hunt together in a group in order to maximise the chance of finding and bringing down prey. In this way, larger prey animals may be hunted, providing more food energy. For example, Harris hawks hunt rabbits in teams.

Advantages of cooperative hunting

1. Co-operative hunting may benefit subordinate animals as well as dominant ones, as they may gain more food than by foraging alone.
2. Less energy is used per individual when hunting as a group.
3. Enables larger prey to be caught.
4. Increases the chance of hunting success.

Social defence

- Vigilance – defence strategies among social animals, such as musk oxen and meerkats, increase the chance of survival. This is because animals can watch for predators while others forage for food. Vigilance behaviour can be seen in prairie dogs and barnacle geese while grazing. In a large flock of barnacle geese, individuals may benefit from an earlier warning of the approach of a predator as there are more eyes to spot their approach.
- Herd movement – a herd of animals will keep close together while travelling, as there is 'safety in numbers'. Females and young are positioned towards the centre of the herd for protection, with watchful males on the perimeter. Groups of musk oxen adopt specialised formations when under attack. Young will be protected in the centre and surrounded by others facing outwards in a circular defensive formation.

Prairie dogs

Altruism and kin selection (influence on survival)

- Altruism – this is a behaviour that benefits the survival chances of other members of a social group, at a cost to the individual. In social animals, such as vampire bats, donor individuals that have had a successful blood meal may help recipient hungry bats by regurgitating and sharing the meal. The recipient benefits by gaining resources but the role of donor and recipient will be reversed in the future. This is called **reciprocal altruism**.
- Kin selection – this is altruistic behaviour that specifically benefits the survival chances of close relatives or kin within a social group. For example, adult chimpanzees giving their food to juvenile members of the social group. This ensures an increased chance of survival of shared genes in the recipient's offspring or future generation.

TOP TIP

Remember: behaviour that appears to be altruistic can be common between a donor and a recipient if they are related (kin).

Social insects

Bees, wasps, ants and termites exist within complex social structures, benefiting the environment by providing 'ecosystem services' such as pollination, decomposition and natural biological control of predator populations.

Social structure in honey bee colonies

Within a bee colony there exist three distinct social levels: queen bee, worker bee and drone. Only the queen and drones contribute reproductively to a colony. Most members of the colony are sterile workers who co-operate with close relatives to raise the offspring of the queen with whom they share genes.

The queen bee is the only bee within the colony that produces eggs. Other female bees do not produce eggs and become worker bees.

Worker bees are sterile and raise relatives to increase survival of shared genes. Instead of reproducing they have roles that include defending the hive, collecting pollen and carrying out waggle dances in order to communicate the direction of food.

The number 15 sticker seen attached to a bee in the image above is used to identify the queen in this hive.

Primate behaviour

Primates, which include humans, monkeys, chimpanzees and gorillas, are organised into a social hierarchy that determines the access rights of an individual to food, mates and available habitat. Complex social behaviours support the social hierarchy.

Behaviour	Feature	Advantage
Parental care	Long period of time	Allows learning of complex social behaviours from parents
Ritualistic displays	Reinforces a leader's position within a group, for example use of vocal sounds and chest beating	Reduces conflict
Appeasement (submissive) behaviour	Use of facial expressions, sexual presentation and lower body posture to signal 'standing down'	Reduces conflict and potentially fatal injury
Alliances with others	Social bonds develop between individuals	Increases social status within the group Alliances are strengthened by grooming

Orangutan social group

Quick Test 28

1. State the advantages of a social hierarchy.
2. Explain the benefit of cooperative hunting to a group of animals.
3. Give one example of a social defence mechanism against predation.
4. State the term used to describe the behaviour of an individual that is of survival benefit to others.
5. State the benefit of forming an alliance within a primate group.

Biodiversity

Biodiversity is the term used to describe the millions of different species of animals, plants and micro-organisms on Earth.

Measuring biodiversity

The three measurable components of biodiversity are:

- genetic diversity
- species diversity
- ecosystem diversity.

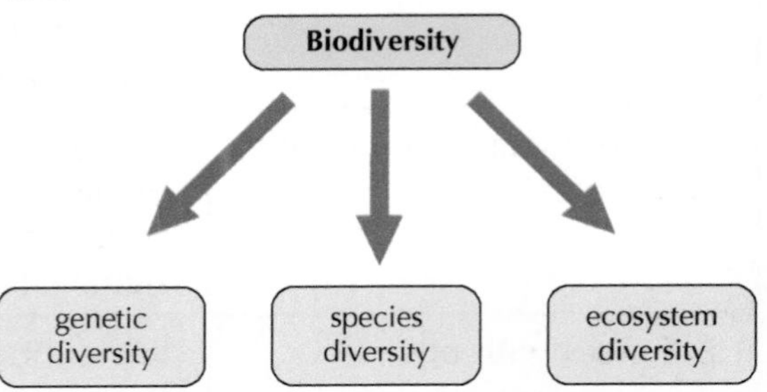

Measurable component	Definition	Importance
Genetic diversity	Genetic diversity comprises the genetic variation represented by the number and frequency of alleles in a population.	Indicates the total number of genetic characteristics of a species. Low genetic diversity makes the species vulnerable to changing climatic conditions due to an inability to adapt.
Species diversity	Species richness = the number of different species present in an ecosystem. Relative abundance = the proportion of each species in an ecosystem.	A high number of species results in a stable ecosystem. If one species dominates (high abundance), species diversity is reduced.
Ecosystem diversity	The number of distinct ecosystems within a defined area.	A region with a wide variety of ecosystems will have greater species diversity. The more remote and isolated an ecosystem is, such as an isolated island, the lower its species diversity.

TOP TIP

If one population of a species dies out, then the species may have lost some of its genetic diversity and this may limit its ability to adapt to changing conditions.

TOP TIP

Remember: a community with one dominant species, such as false oat-grass, has a lower species diversity than one with the same number of species (species richness) but no particularly dominant species.

Threats to biodiversity

Threats to biodiversity include:

1. exploitation
2. the bottleneck effect
3. habitat loss
4. introduced, naturalised and invasive species.

TOP TIP

'Viable' means capable of surviving under certain environmental conditions.

Exploitation and recovery

Overexploitation resulting from excessive hunting and fishing by humans has threatened the survival of many species. The North Sea cod populations are threatened by continued removal of fish that has prevented population recovery. Current population numbers are barely sustainable.

The northern elephant seal found in the coastal waters of California and Mexico became almost extinct in the late 1800s due to overhunting.

The species was reduced to a population of 20–40 individuals but remarkably managed to recover to current numbers of approximately 100,000. The lasting legacy is a **viable** elephant seal population that has a low level of genetic diversity. In the future, this may be detrimental to the species as they may lack the ability to adapt if environmental conditions change.

Bottleneck effect

The **bottleneck effect** is a sharp reduction in a population of a species for at least one generation. The reduction in numbers could be due to overexploitation such as overhunting or due to a natural disaster such as a tsunami or an earthquake that wipes out most of the population.

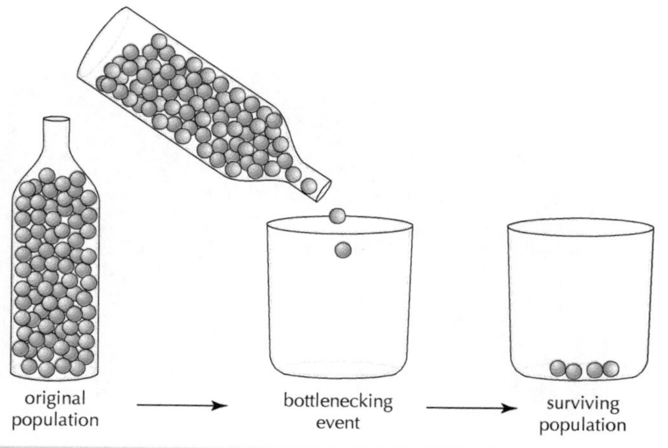

TOP TIP

Following a bottleneck the surviving small population can lack genetic diversity. This combined with interbreeding can result in poor reproductive rates.

The bottleneck results in a change in the frequency of alleles in a population and a decrease in genetic variation.

The surviving small population may lose the genetic diversity necessary to enable evolutionary responses to environmental change. This can be critical for many species as inbreeding can result in poor reproductive rates.

Cheetahs experienced bottleneck events due to climatic conditions, during the last ice age and more recently due to overhunting. As a result, modern cheetahs have low genetic diversity, are prone to disease and have poor reproductive rates due to inbreeding.

Habitat loss and fragmentation

The clearing of habitats has led to habitat fragmentation. The resulting increase in competition for limited resources decreases biodiversity as some species outcompete others for resources. As a result, isolated small fragments exhibit a lower species diversity. The edges of the habitat fragments erode and species living on the edges (edge species) may compete for limited resources with those living in the middle of the fragment.

Remedy for habitat fragmentation

Habitat corridors connect fragmented pieces of a habitat. They can be natural or man-made, and allow the movement of animals between fragments, increasing access to food and choice of mate. Linking larger fragments with isolated fragments may lead to recolonisation after local **extinctions**.

Introduced, natural and invasive species

An exotic species, sometimes called an introduced or alien or non-native species may be introduced from a different geographical location with potentially devastating effects. The Chinese mitten crab was introduced to Britain as an exotic species several years ago, and has now naturalised, becoming an **invasive** species moving through waterways from England to Scotland. Salmon stocks in the River Tweed may be threatened by the invasion of the Chinese mitten crab currently found in the River Clyde, several miles away. Mitten crabs can travel across land and, if they enter the River Tweed, they will aggressively compete with salmon for available food. This may mean that salmon move back out to sea, leaving the river ecosystem and affecting the local economy, which depends upon salmon fishing.

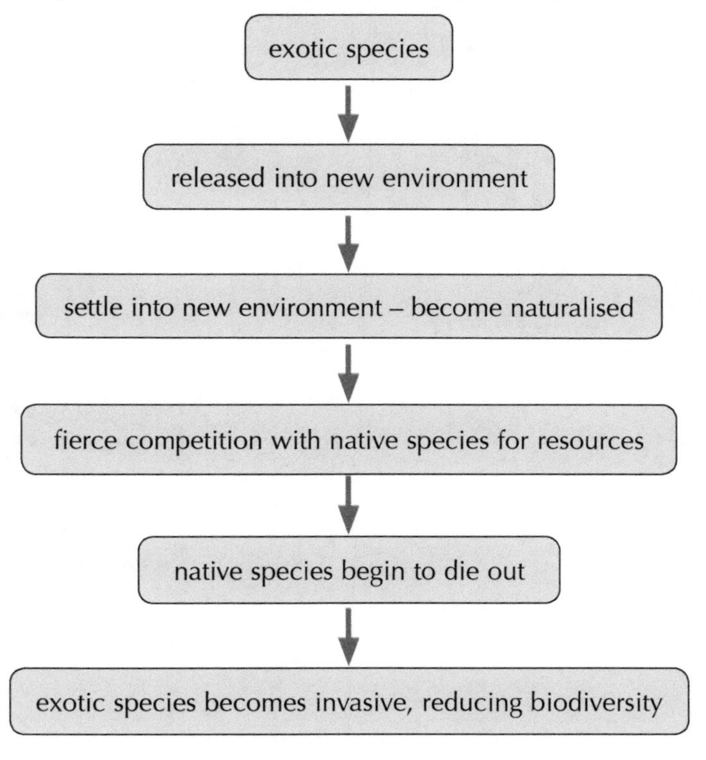

Remember the definitions:

Introduced species (non-native)	Species that humans have moved intentionally or accidentally to new geographic locations.
Naturalised species	Species that have become established within wild communities.
Invasive species	These are species that spread rapidly and eliminate native species, thereby reducing species diversity.

Invasive species are successful because:

1. they are free of predators, parasites, competitors and pathogens that limit their population size in their native habitat
2. they prey on native species, outcompete with them for resources or hybridise with them.

Spanish bluebell
Hyacinthoides hispanica

Hybrid bluebell
Hyacinthoides x massartiana

Native British bluebell
Hyacinthoides non-scripta

TOP TIP

Hybridisation is the process of breeding two different species to produce an organism called a hybrid. For example, the Spanish bluebell and the native British bluebell can reproduce to make a hybrid.

Quick Test 29

1. Define the term 'biodiversity'.
2. State the three measurable components of biodiversity.
3. Identify three different threats to biodiversity.
4. How can the problem of habitat fragmentation be addressed?
5. Explain what is meant by the term 'invasive species'.

Glossary

absorption spectrum: graph showing the different wavelengths of light absorbed by photosynthetic pigments

acetyl-CoA: formed from an acetyl group and co-enzyme A during cell respiration if oxygen is present

action spectrum: graph showing the rate of photosynthesis at each wavelength of light

activation energy: the energy needed to start a reaction

active site: area of the enzyme which binds to the substrate

aestivation: state of inactivity and low metabolism entered into by animals in response to high temperatures and drought

allopatric speciation: occurs when populations become isolated by geographical barriers that prevent gene flow

amplification: refers to the synthesis of multiple copies of a DNA fragment using the polymerase chain reaction (PCR) technique

antiparallel: two strands of the DNA molecule run in opposite directions from carbon 3' (deoxyribose end) to carbon 5' (phosphate end)

artificial chromosome: chromosome made in the lab and used as vector to transfer larger fragments of foreign DNA

aseptic: absence of all micro-organisms except the one being cultured, as a result of sterilisation procedures

ATP synthase: an enzyme that catalyses the synthesis of ATP from ADP and inorganic phosphate

bioaccumulation: build-up of a chemical, such as a pesticide, in the body of an organism

bioinformatics: the use of computers and statistical analysis to compare sequence data

biomagnification: increasing concentration of a chemical at each trophic level in a food chain

biosynthesis: the building of complex molecules by a living cell from basic subunits such as amino acids joining together to make proteins

bottleneck effect: loss of genetic diversity in a species due to an event such as a tsunami

Calvin cycle: second stage of photosynthesis in which carbon is fixed

carotenoids: pigments that extend the wavelengths of light absorbed and pass the absorbed energy to chlorophyll for photosynthesis

cell organelle: small membrane-bound structure within the cytoplasm of an animal or plant cell

chloroplast: cell organelle containing small circular chromosomes. Site of photosynthesis

chromosomes: thread like structures found in the nucleus of a cell. They are composed of DNA tightly coiled and packaged around associated proteins called histones

citrate: intermediate compound in the citric acid cycle

citric acid cycle: cyclical series of reactions operating under aerobic conditions in the matrix of a mitochondrion

competitive inhibitor: substance that binds reversibly to the active site of an enzyme, reducing the rate of reaction

consequential dormancy: when an organism goes into a state of dormancy after adverse conditions develop

cultivar: a plant that has been selected and purposely grown for use in agriculture and horticulture

daily torpor: a period of reduced activity in some animals with high metabolic rates

death phase: stage when cells are dying due to toxic accumulation of metabolites or the lack of nutrients in culture

dehydrogenase: enzymes that catalyse the removal of hydrogen ions and electrons from a substrate

deletion (chromosome mutation): a section of a chromosome in the middle or end of a chromosome is deleted

deletion (single gene mutation): a single gene mutation resulting in the deletion of a nucleotide in a DNA sequence, which results in a frame-shift mutation

differentiation: a process by which cells express genes that produce proteins for that specialised cell type

direct contact: mechanism by which parasites are transmitted to new host

directional selection: mode of natural selection in which one extreme in a phenotype range is favoured

disruptive selection: mode of natural selection in which two or more phenotypes are selected for

DNA polymerase: enzyme that adds DNA nucleotides to the 3' end of the primer and newly forming strand

dormancy: a period of inactivity in the life cycle of a plant or animal where growth almost completely stops and essential metabolic processes only are maintained to sustain life

double helix: 3D shape of a DNA molecule discovered by Watson and Crick

duplication: a selection of a chromosome is added from its homologous partner; it allows potential beneficial mutations to occur in a duplicated gene while the original gene can still be expressed

effectors: tissue that brings about a corrective response as a result of a nerve impulse

electron transport chain: series of carrier proteins attached to the inner membrane of the mitochondrion

ethanol: an alcohol produced by yeasts during fermentation of sugars

eukaryotic cell: a cell such as a plant, animal or fungal cell that contains membrane-bound organelles such as a nucleus and mitochondria

exon: coding region of DNA

exponential (log) phase: period when cell numbers double for each growth cycle

extinction: the irreversible loss of a whole species or group of organisms

feedback inhibition: occurs when the end product in a metabolic pathway reaches a critical concentration and inhibits an earlier enzyme in the pathway preventing further synthesis of end-product

fermentation: breakdown of glucose in the absence of oxygen

fermenter: large container used to grow micro-organisms on a commercial scale

field trial: an experimental investigation to test the growth of either new cultivars or genetically modified plants in real growth conditions

food security: the ability of human populations to access food of sufficient quality and quantity

food supply: availability of food

frame-shift mutation: a nucleotide deletion or insertion results in all the codons and amino acids after the mutation being changed

genetic code: the base sequence of DNA

gene expression: process by which a base sequence in a gene is used to make a functional gene product

generation time: time taken for one cell to become two

genetically modified crop: plant that has undergone an alteration in its DNA allowing it to code for a new, beneficial protein

genome: the entire hereditary information encoded in DNA

genotype: the alleles or 'forms of a gene' an organism possesses

glycolysis: initial series of reactions of cellular respiration that take place in the cytoplasm with or without the presence of oxygen

hibernate: state of low metabolism and body temperature entered into by animals during low temperatures in order to ensure survival

histone: a protein molecule around which DNA is tightly coiled within the nucleus of a cell

homeostasis: general term for the maintenance of the internal environment using a variety of mechanisms

homologous: refers to a pair of chromosomes that are similar in length and gene position

horizontal gene transfer: where genes are transferred between individuals in the same generation

hybrid vigour: offspring from a cross of genetically different parents showing increased growth rate or resistance to disease

hypothalamus: region of the brain associated with monitoring temperature

inbreeding: breeding of selected plants or animals for several generations to eliminate heterozygotes

inbreeding depression: loss of heterozygositiy due to continued selective breeding

induced fit: the ability of an enzyme to alter slightly the 3D shape of the active site in order to better fit the substrate after the substrate binds

inhibitor: a molecule that reduces the rate of an enzyme catalysed reaction, or stops the reaction completely

innate behaviour: inherited pattern of behaviour that has not been learned

insertion mutation: a single gene mutation that involves the insertion of a nucleotide into a DNA sequence that results in a frame-shift mutation

integrated pest management: system using a combination of pest-control methods, cultural, chemical and biological, to control size of a pest population

internal environment: body cells and the tissue fluid that surrounds them

intron: non-coding region of DNA

invasive: a species introduced from another ecosystem that outcompetes aggressively with the native species for resources, eventually taking over and eliminating the native species

inversion mutation: mutation in which a section of a chromosome is reversed

lactate: compound formed as an end-product of fermentation during strenuous exercise

lagging strand: DNA strand replicated discontinuously in fragments

lag phase: period when enzymes are induced to metabolise substrates

leading strand: DNA template strand replicated continuously from the 3′ to the 5′ end

learned behaviour: animal exhibits response to a situation as a result of previous experience

ligase: enzyme used to join fragments of DNA together during DNA replication and enzyme used to seal the gene into a plasmid in recombinant DNA technology

mature mRNA transcript: strand of messenger RNA that has had all introns removed and exons spliced together

metabolic pathway: an integrated and controlled pathway of enzyme-catalysed reactions within a cell

metabolic rate: a measure of the rate of energy used in a given time

metabolism: all the chemical reactions that occur within a living organism

migration: the movement of populations of animals to a more favourable environment, in order to avoid adverse conditions

missense mutation: a base is substituted for another resulting in a single amino acid change in the sequence or no change in the amino acid sequence

mitochondrion: cell organelle within the cytoplasm of eukaryotic cells that generates ATP through the processes of the citric acid cycle and electron transport chain in cell respiration

model organism: a species that is easy to grow and maintain in a laboratory and whose biology has been widely studied

mutagenesis: a process where the genetic information of an organism is altered in a stable manner resulting in a mutation

mutation: a change in the genetic composition of a cell

multipotent: term used to describe the property of tissue stem cells that allows differentiation into all the types of cell found in a particular tissue type

mutant: organism resulting from an irreversible change in genotype that causes a subsequent change in phenotype (physical characteristics)

negative feedback control: type of corrective mechanism for restoring and maintaining the dynamic state of an organism's internal environment in which a departure from a set value for a variable is detected and a response made to reduce the intensity of the increasing stimulus

non-competitive inhibitor: substance that binds to an area other than the active site of an enzyme causing a change in the shape of the active site so the substrate no longer fits

nonsense mutation: a substitution of one base for another resulting in a premature stop codon being produced

orientation: the position in space of the reactants in an enzyme-catalysed reaction to ensure the reaction proceeds

origin of replication: particular part of the DNA molecule where replication is started. Found in plasmids and circular chromosomes

oxaloacetate: intermediate compound that joins with acetyl to form citrate

peptide bond: type of bond formed between amino acids

personalised medicine: the use of genome information to predict the likelihood of developing certain diseases and to select the most effect drugs and dosage to treat a disease

pharmacogenetics: the use of genome information to decide on choice of drug treatment

phenotype: physical characteristics of an organism, determined by the genotype

phosphorylation: addition of a phosphate group to a molecule

photolysis: process occuring in the chloroplast that results in a water molecule being split into hydrogen and oxygen

phylogenetics: the study of relationships and evolutionary history

pigments: chemicals found in the chloroplast that can absorb light of different wavelengths

plasmid: small circular piece of DNA found in bacterial and yeast cells

pluripotent: term used to describe cells in the early embryo that can be grown in the laboratory and have the potential to differentiate into any type of cell

polymerase chain reaction (PCR): method by which multiple copies of a DNA fragment can be made in the laboratory; PCR is used to amplify a DNA fragment so that enough genetic material can be analysed to help solve crimes, help settle paternity suits and help diagnose genetic disorders

polypeptide: chain of many amino acids linked together by peptide bonds

predictive dormancy: when an organism goes into a state of dormancy before adverse conditions develop

primary mRNA transcript: single strand of messenger RNA formed during transcription, which contains both introns and exons

primer: a short sequence of nucleotides that are complementary to specific target sequences at the two ends of a region of DNA to be amplified

prokaryotic cell: cell, such as a bacterium, which does not contain membrane-bound organelles

protein pore: a channel through the membrane, allowing the free passage of any molecule of the appropriate size

protein pump: protein embedded within the phospholipid layers that can transport molecules across the cell membrane against the concentration gradient requiring energy

protein synthesis (gene expression): the name given to the process where the DNA base sequence on a gene is converted to an amino acid sequence, through the processes of transcription and translation – this is called 'gene expression'

pyruvate: formed from the splitting of a glucose molecule during the first stage of cell respiration called glycolysis; this occurs in the cytoplasm of the cell, with or without the presence of oxygen

receptor: specialised group of cells that can detect changes in an organism's body or in the environment

reciprocal altruism: when the cost of helping another individual within a social group is repaid

recombinant DNA technology: manipulation of genetic material, often by transferring DNA between different organisms

regulatory gene sequence: DNA sequence that controls gene expression

replication fork: point at which hydrogen bonds between complementary bases of DNA template strands break, causing separation during DNA replication

resistant stage: term used to describe a stage in the transmission of a parasite that is capable of withstanding adverse environmental conditions until it comes into contact with a new host

respirometer: apparatus used to measure the metabolic rate of an organism by measuring the volume of oxygen uptake per unit of time

restriction endonuclease: enzyme that cuts DNA into fragments only at a restriction site

restriction site: short sequence found on both strands of a DNA molecule that can be recognised by an endonuclease

ribosomes: structures found in the cytoplasm where a mature mRNA transcript is translated into a polypeptide chain

RNA polymerase: the enzyme that controls the synthesis of the mRNA strand during the transcription process in protein synthesis; it adds RNA nucleotides onto the 3′ end of the newly forming mRNA strand

RNA splicing: process by which mature RNA transcript is formed; introns are removed and exons are spliced together

RuBisCo: enzyme that attaches carbon dioxide to RuBP

RuBP: intermediate metabolite in the Calvin cycle that accepts carbon dioxide molecule

selectable marker: section of DNA that allows genetically modified host cell to be identified

self-renewal: ability of stem cells to generate new stem cells

set point: indicates that the internal environment is stable at the correct point within a negative feedback pathway

speciation: the evolution of new species through natural selection, as a result of a parent population being split into two; each sub-population is exposed to different selection pressures, and evolves down different pathways; eventually, the two sub-populations cannot interbreed, proving that speciation has occurred

species: group of organisms capable of interbreeding and producing fertile offspring, and which does not normally breed with other groups

splice-site: the boundary between an exon and an intron

stabilising selection: mode of selection in which the average phenotype is selected for and extremes of the phenotype are selected against

stationary phase: period when growth and death phases balance each other out

stem cell: an undifferentiated cell with no specific function, which has the potential to differentiate into different types of specialised cells

sticky ends: short lengths of single-stranded DNA left exposed after the action of restriction endonucleases

substitution mutation: type of gene mutation in which one nucleotide is substituted for another in a DNA sequence; there are three types of substitution mutation: missense, nonsense and splice-site

sympatric speciation: speciation that occurs when populations become isolated by ecological and behavioural barriers that prevent gene flow

template strand: single strand of double-stranded DNA molecule that acts as a template for DNA replication

thermocycling: stages during the polymerase chain reaction process where a DNA fragment is exposed to a sequence of different temperatures; first 92–98°C, then 50–65°C and finally 72–80°C

total cell count: a measure of all the cells in a culture, both living and dead

transcription: first stage in protein synthesis (gene expression) that occurs in the nucleus of the cell and results in the production of a strand of mRNA called the 'primary transcript'

transition state: during an enzyme-catalysed reaction, the point at which the reaction can proceed and products are formed

translation: the second stage in gene expression where the mature mRNA transcript is translated into an amino acid sequence using tRNA molecules

translocation: a chromosome mutation resulting from the addition of a section of a chromosome to another chromosome that is not its homologous partner

vector: plasmid or artificial chromosome that carries a gene or genes for insertion into another organism

vector organism: an organism that transports a parasite from one host to another

vegetative reproduction: any type of asexual reproduction in plants

vertical gene transfer: the transfer of genes from parents to offspring in the next generation

viable (population): a species that is capable of surviving and reproducing

viable cell count: a measure of only the living cells in a culture

Answers to Quick Tests

Quick Test 1

1. Deoxyribose sugar and phosphate.
2. The two strands of DNA run in opposite directions to each other (deoxyribose 3′ end and phosphate 5′ end).
3. The genetic code is the base sequence of DNA.

Quick Test 2

1. A prokaryote has a large circular chromosome in the cytoplasm and a eukaryote has linear chromosomes in the nucleus.
2. Plasmid.
3. Circular chromosome.

Quick Test 3

1. A primer allows DNA replication to start as it binds to the DNA template strand and in doing so allows DNA polymerase to add DNA nucleotides.
2. 3′ end of primer
3. Ligase.
4. DNA polymerase adds nucleotides, to the 3′ end of the primer and newly forming strand, in one direction only, resulting in the lagging strand being replicated in fragments.

Quick Test 4

1. The enzyme can withstand the high temperatures of PCR.
2. DNA amplified by PCR can be used to help solve crimes, help settle paternity suits, and be used to diagnose genetic diseases.
3. Gel electrophoresis is used to separate macromolecules; for example, DNA fragments.

Quick Test 5

1. DNA contains deoxyribose sugar and RNA contains ribose sugar. DNA contains the base thymine whereas RNA contains the base uracil.
2. The proteins produced as a result of gene expression.
3. Ribose sugar, phosphate and a base.
4. A ribosome is made up of rRNA and protein.
5. Exons are coding regions and introns are non-coding regions.

Quick Test 6

1. mRNA sequence – UUC GCA CCA UAC UGG
2. RNA polymerase moves along the DNA unwinding the double helix and breaking the hydrogen bonds between the bases. RNA polymerase synthesises a primary transcript of mRNA from RNA nucleotides by complementary base pairing.
3. The introns, non-coding regions, of the primary mRNA transcript are removed and the exons, coding regions, are joined together in the mature transcript.
4. A ribosome

Quick Test 7

1. An anticodon is a triplet of bases on a tRNA molecule.
2. tRNA molecules are specific to, and carry, only one aminoacid depending on their anticodon.
3. Peptide bond
4. A stop codon is a triplet of bases on mRNA that stops translation.

Quick Test 8

1. Exons
2. 496. Working out: 2976/2 = 1488 nucleotides in 1 strand. 1488/3bases = 496 amino acids (remember a gene is single strand and three bases (codon) = 1 amino acid).
3. Alternative RNA splicing. Different exons are retained/included from the primary transcript, resulting in a different sequence of exons in the mature mRNA transcript.

Quick Test 9

1. Cellular differentiation is the process by which a cell expresses certain genes to produce proteins characteristic of that cell type. This allows a cell to carry out specialised functions.
2. All the genes in embryonic stem cells can be switched on and expressed so these cells can differentiate into any type of cell.
3. Blood stem cells.
4. Human stem cells can be used in research to study how diseases develop/or be used as model cells for the testing of drugs in the laboratory.

Quick Test 10

1. The genome is made up of genes and other DNA sequences that do not code for proteins.
2. Some noncoding regions regulate transcription and others are transcribed but never translated.
3. A single gene mutation involves the alteration of a DNA nucleotide sequence as a result of the substitution, insertion or deletion of nucleotides.
4. A frame-shift mutation causes all of the codons and all the amino acids after the mutation to be changed.
5. Splice-site mutations result in some introns being retained and/or some exons not being included in the mature transcript.

Quick Test 11

1. Down's syndrome.
2. Deletion, duplication, translocation and inversion.
3. Translocation.
4. Gene duplication results in an extra copy of a gene that may randomly mutate to form an allele that confers a survival advantage.

Quick Test 12

1. Faster evolutionary change (within same generation) in prokaryotes in comparison to eukaryotes.
2. Unable to find food due to high levels of competition for limited resources. Lack of DNA sequences that increase survival.
3. In eukaryotes, vertical gene transfer involves the transfer of genes from parent to offspring in another generation via sexual or asexual reproduction.
4. Isolating barriers prevent gene flow between sub-populations.
5. Sympatric speciation involves behavioural and ecological barriers.

Quick Test 13

1. It allows analysis and comparison of the entire DNA sequence between one organism and another.
2. Bioinformatics combines computer and statistical analysis in order to compare sequence data between individuals and between species.
3. Sequence data and fossil evidence.
4. A constant rate of mutation.
5. Bacteria, archaea and eukaryotes.

Quick Test 14

1. Anabolic or catabolic.
2. Metabolite 1 may be converted to metabolite 2 and back again. The cell has control over the production and use of metabolite 2.
3. Carrier protein on cell membrane that moves molecules across against the concentration gradient.
4. Protein pores contain a channel for the transport of large molecules across the plasma membrane.
5. Active transport.

Quick Test 15

1. Lowering activation energy speeds up the rate of a chemical reaction.
2. Induced fit describes the ability of the active site of an enzyme to alter shape slightly to accommodate the substrate molecule.
3. The rate of reaction increases until all active sites are full, then it remains constant.
4. Orientation of substrate molecules within the active site lowers activation energy.
5. Slows down or stops an enzyme-catalysed reaction.
6. Non-competitive inhibitor.
7. Increasing the concentration of the substrate.

Quick Test 16

1. Muscle contraction, cell division, replication of DNA, transport, transmission of electrical impulses in nerves, synthesis of new molecules.
2. ADP + P_i to ATP
3. ATP is constantly being broken down and regenerated, very little is stored by the body.
4. NAD – nicotinamide adenine dinucleotide.
5. Pyruvate.
6. Cytoplasm.
7. 2 ATP needed to break glucose molecule, 4 ATP generated from broken glucose bonds, net gain to cell of 2ATP/more ATP generated in the pay-off phase in comparison to investment phase, leading to a net gain.

Quick Test 17

1. Matrix of mitochondrion.
2. NAD.
3. Electron transport chain.
4. ATP synthase is a protein molecule rotated by the flow of electrons, resulting in the phosphorylation of ADP + P_i to ATP.
5. Oxygen.

Quick Test 18

1. The rate at which an organism releases energy during the process of respiration in a given time.
2. Oxygen consumption, carbon dioxide produced, energy released as heat.
3. In a mammalian heart, oxygenated blood is kept completely separate from deoxygenated blood and is pumped at higher pressure than in fish. In a fish heart, oxygenated and deoxygenated bloods mix.
4. 3 heart chambers (2 atria and 1 ventricle).

Quick Test 19

1. Energy costs to the organisms are low.
2. Conformers are less able to adapt to and survive changes to the external environment.
3. Control of blood water, blood glucose, body temperature.
4. High energy cost to organism.
5. Change detected by receptor cells, signal sent to effector organ, change brought about, system restored to set point.
6. It is the temperature monitoring centre.
7. Sweat glands, blood vessels of the skin, hair erector muscles, skeletal muscles.

Quick Test 20

1. Consequential dormancy.
2. Reduced metabolic rate conserves energy.
3. Migration.
4. Innate behaviour is inherited, learned behaviour occurs as a result of trial and error experiences.
5. Satellite tracking can be done enitirely electronically whereas ringing can only be tracked manually.

Quick Test 21

1. Substrate concentration, temperature, pH and oxygen concentration.
2. Amino acids and fatty acids.
3. Generation time is the time taken for one cell to divide into two daughter cells.
4. During the exponential phase, cells divide at their maximum rate.
5. During the stationary phase, the number of micro-organisms dying is balanced by the number of micro-organisms becing produced.

Quick Test 22

1. Mutagenesis or recombinant DNA technology.
2. Plasmid.
3. Restriction endonuclease 'cuts' DNA at a restriction site and is used to open a plasmid and cut out a desired gene from the DNA of another species.
4. The presence of a marker gene confirms that a bacterial cell successfully contains a transferred gene within the plasmid.

Quick Test 23

1. Food security – the ability of human populations to access food of sufficient quality and quantity.
2. Grow high-yielding cultivar, use fertilisers, reduce pests and diseases, reduce competition between plants, use optimal seeding density.
3. Light intensity, carbon dioxide concentration, temperature.
4. Much energy is lost between trophic levels; therefore, eating plants directly saves energy.

Quick Test 24

1. Chlorophyll a and b.
2. Absorption spectrum – shows the wavelengths of light absorbed by each pigment.

   Action spectrum – shows how good each wavelength of light affects the relative rate of photosynthesis.
3. ATP and NADPH.
4. RuBisCo.
5. Production of glucose and regeneration of RuBP.
6. NADP.

Quick Test 25

1. Field trials can be used to compare the performance of different cultivars or treatments and to evaluate genetically modified crops.
2. 'Inbreeding depression' occurs when related members of the same species are interbred for many generations, and this results in undesirable recessive alleles being expressed in the phenotypes of the offspring.
3. Crossbreeding animals results in offspring with hybrid vigour.
4. F_1 hybrids don't usually generate much variation in the F_2 generation.
5. Analysis of a genome sequence allows alleles that express desirable phenotypes to be selected and used in future breeding programmes. Single genes for desirable characteristics can also be inserted into the genomes of crops to create GM cultivars with improved characteristics.

Quick Test 26

1. Weeds, pests, diseases.
2. Annual weeds – long-term seed viability, rapid growth/life cycle, produce many seeds.

   Perennial weeds – vegetative reproduction, food reserves in storage organs.
3. Nematode worms, molluscs, insects.

4. Crop rotation – breaks pest life cycle. Ploughing – destroys perennial weeds.
   Optimal seeding density reduces competition and prevents establishment of weeds.
5. Systemic and selective.
6. Less negative impact on the environment; chemical use is either minimised or eliminated.

Quick Test 27

1. Ethically desirable for animal welfare; increases productivity if animals are in better living conditions/less stressed; increases rate of reproduction.
2. (a) stereotypy; misdirected behaviour; failed sexual/parental behaviour; altered levels of activity.
   (b) misdirected behaviour.

Quick Test 28

1. Reduction in conflict. Increased chance of the dominant allele's favourable genes being passed on to offspring.
2. Less energy is used per individual. Greater chance of hunting success. Subordinate as well as dominant animals may gain more food than by foraging alone.
3. Specialised formation when under attack, thus protecting their young.
4. Altruistic.
5. Increases social status within the group.

Quick Test 29

1. 'Biodiversity' refers to the total number of different species of animal, plant and microorganism on Earth.
2. Genetic diversity, species diversity and ecosystem diversity.
3. Exploitation, habitat fragmentation, introduced species that become invasive and the bottleneck effect.
4. Formation of a habitat corridor.
5. An invasive species spreads rapidly and eliminates native species, thereby reducing species diversity.

Higher BIOLOGY

For SQA 2019 and beyond

Practice Papers

John Di Mambro, Deirdre McCarthy and Stuart White

Revision advice

The need to work out a plan for regular and methodical revision is obvious. If you leave things to the last minute, it may result in panic and stress, which will inhibit you from performing to your maximum ability. If you need help, it is best to find this out when there is time to put it right. Revision planners are highly individual and you need to produce one that suits you. Use an area of your home that is set aside only for studying if possible, so that you form a positive link and in this way will be less liable to distractions. Not only do you need a plan for revising Higher Biology, but also for all your subjects. Below is one revision plan but you will have your own ideas here!

Work out a revision timetable for each week's work in advance – remember to cover all of your subjects and to leave time for homework and breaks. For example:

Day	***6–6.45pm***	***7–8pm***	***8.15–9pm***	***9.15–10pm***
Monday	Homework	Homework	English revision	Biology revision
Tuesday	Maths revision	Physics revision	Homework	Free
Wednesday	Geography revision	English revision	Biology revision	Maths revision
Thursday	Homework	Physics revision	Geography revision	Free
Friday	English revision	Biology revision	Free	Free
Saturday	Free	Free	Free	Free
Sunday	Maths revision	Physics revision	Geography revision	Homework

Make sure that you have at least one evening free each week to relax, socialise and re-charge your batteries. It also gives your brain a chance to process the information that you have been feeding it all week.

Arrange your study time into sessions that suit you, with a 15-minute break in between. Try to start studying as early as possible in the evening, when your brain is still alert, and be aware that the longer you put off starting, the harder it will be.

If you miss a session, do not panic. Log this and make it up as soon as possible. Do not get behind in your schedule – discipline is everything in being a successful student.

Study a different subject in each session, except for the day before an exam.

Do something different during your breaks between study sessions – have a cup of tea, or listen to some music. Do not let your 15 minutes expand into 20 or 25 minutes!

Have your class notes and any textbooks available for your revision to hand, as well as plenty of blank paper, a pen, etc. If relevant, you may wish to have access to the internet, but be careful you restrict using this only for supporting your revision. You may also like to make keyword sheets like the example below:

Keyword	***Meaning***
Ribosome	Structure in cell that is the site of protein synthesis
Fungicide	Chemical that kills fungi

Flashcards are another excellent way of practising terms and definitions. You can make these easily or buy them very cheaply. Use flashcards either to recall the keyword when you see the meaning or to give the meaning when you see the keyword. There are several websites that are free to use and give you the ability to generate flashcards online. If you collaborate with your friends and take different sections of the course, you can merge these into a very powerful learning and revision aid.

Finally, forget or ignore all or some of the advice in this section if you are happy with your present way of studying. Everyone revises differently, so find a way that works for you!

Links to the syllabus

Key area	P1 – Paper 1 P2 – Paper 2	
	Exam A	**Exam B**
The structure of DNA	P1:2 P2:1	P1:1 P2:2(c)
Replication of DNA	P1:3 P2:2	P1:2 P2:3(a) (b)
Gene expression	P1:4 P2:4	P1:4, 7 P2:4(a) 5A
Cellular differentiation	P1:6 P2:3	P2:6(a)(b)
The structure of the genome	P1:4	
Mutations	P1:5 P2:8(c)	
Evolution	P1:8	P1:9
Genomic sequencing	P2:5	P1:8 P2:8 19B

Key area	Exam A	Exam B
Metabolic pathways	P1:9, 11	P1:10 P2:9
Cellular respiration	P1:10 P2:6	P1:11
Metabolic rate	P1:12	P1:13 P2:5B
Metabolism in conformers and regulators	P1:13 P2:7A, 9	P1:14 P2:11
Metabolism and adverse conditions		P2:14
Environmental control of metabolism	P1:14	
Genetic control of metabolism	P2:10(a)	P2:16

Key area	***P1 – Paper 1 P2 – Paper 2***	
	Exam A	***Exam B***
Food supply, plant growth and productivity	P1:15, 16 P2:7B, 11(a) (b) (c) (ii)	P1:15 P2:15
Plant and animal breeding	P1:18	P1:17
Crop protection	P1:19 P2:12(c) (d)	P1:18 P2:17
Animal welfare	P2:13(a) (b)	P1:19, 20 P2:19A(i)
Symbiosis	P2:16A	P1:21
Social behaviour	P1:20, 21 P2:13(d) (e)	P1:25 P2:19A(ii)
Components of biodiversity	P1:23 P2:16B(i)	P1:24
Threats to biodiversity	P2:16B(ii)	P1:23

Planning investigations	P1:25 P2:8(b) (ii) 11(c) (i) (iii) 13(c) 14(a) (b)(c)(d)	P1:3 P2:1(c) 3(d) 18(c) (d)
Selecting information	P2:8(b) (i) 14(f) 15(a)	P1:22 P2:6(c) 7(a) (b) 13(a)
Presenting information	P2:8(a) 14(e)	P2:7(c) 18:(a)
Processing information	P1:1, 22, 24 P2:8(c) P2:12(a) 14(g) 15(b)	P1:6, 12, 16 P2:2(a) (b) 2(d) (e) 3(c) 10(a) (b) 12
Predicting and generalising	P1:7	P1:5 P2:1(a) (b) 6(d)
Concluding and explaining	P1:17 P2:8(b) 14(h)	P2:4(b) 10(c) 18(b)
Evaluating	P2:8(b) (iii)	P2:13(b)

PAPER 1 ANSWER GRID

Mark the correct answer as shown ◉

	A	B	C	D
1	○	○	○	○
2	○	○	○	○
3	○	○	○	○
4	○	○	○	○
5	○	○	○	○
6	○	○	○	○
7	○	○	○	○
8	○	○	○	○
9	○	○	○	○
10	○	○	○	○
11	○	○	○	○
12	○	○	○	○
13	○	○	○	○
14	○	○	○	○
15	○	○	○	○
16	○	○	○	○
17	○	○	○	○
18	○	○	○	○
19	○	○	○	○
20	○	○	○	○
21	○	○	○	○
22	○	○	○	○
23	○	○	○	○
24	○	○	○	○
25	○	○	○	○

Higher Biology

Practice Papers for SQA Exams

Exam A

Fill in these boxes and read what is printed below.

Full name of centre

Town

Forename(s)

Surname

Answer all of the questions in the time allowed.

Total marks – 120

Paper 1 – 25 marks – Duration 40 minutes

Paper 2 – 95 marks – Duration 2 hours 20 minutes

Read all questions carefully before attempting.

Write your answers in the spaces provided, including all of your working.

PAPER 1 – 25 marks

Attempt ALL questions

Answers should be given on the separate answer sheet provided.

1. A length of double-stranded DNA is 250 base pairs long. Of these bases, 60 are cytosine.

The percentage of thymine bases present in this part of the DNA is

A 12%

B 19%

C 24%

D 38%

2. In which form does the genetic material of a eukaryote exist within the nucleus?

A Linear chromosomes

B Circular chromosomes

C Linear plasmids

D Circular plasmids.

3. The following are steps in the process of polymerase chain reaction (PCR).

1 Cooling allows primers to bind to target sequences

2 DNA is heated to separate the strands

3 Repeated cycles of heating and cooling amplify this region of DNA

4 Heat-tolerant DNA polymerase then replicates the region of DNA.

Choose the letter that puts the above steps into the correct order.

A $1 \rightarrow 2 \rightarrow 3 \rightarrow 4$

B $2 \rightarrow 1 \rightarrow 3 \rightarrow 4$

C $2 \rightarrow 1 \rightarrow 4 \rightarrow 3$

D $4 \rightarrow 2 \rightarrow 3 \rightarrow 1$

4. Which of the following correctly describes the structure of the genome of an organism:

A very little of a eukaryotic genome consists of non-coding sequences

B it is made up of RNA sequences and non-coding DNA

C the entire hereditary encoded in DNA

D it is made up of RNA sequences and coding DNA?

5. The following sequence of bases codes for three amino acids.

T–G–C–A–A–G–C–G–T

The sequence of bases is altered by a mutation and is changed to the sequence below.

T–G–C–A–A–C–C–G–T

Which type of mutation has occurred?

A Deletion

B Insertion

C Substitution

D Duplication.

6. Which of the following statements **correctly** describes cells and their potential to differentiate?

A Stem cells in plants are pluripotent

B All the genes in tissue stem cells can be switched on

C Tissue stem cells are pluripotent

D Blood stem cells located in bone marrow can give rise to all types of blood cell.

7. The graph on the right shows the rate of photosynthesis at two different levels of carbon dioxide concentration at 20°C.

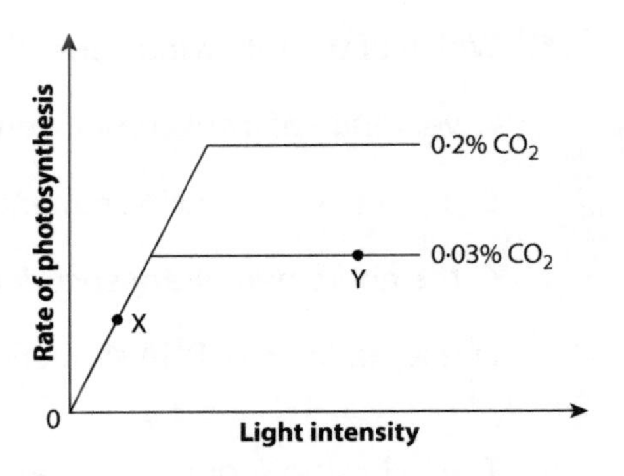

From the evidence given, which line in the table below identifies the factors most likely to be limiting the rate of photosynthesis at points **X** and **Y** on the graph?

	X	*Y*
A	light intensity	CO_2 concentration
B	temperature	light intensity
C	CO_2 concentration	temperature
D	light intensity	temperature

8. Which of the following terms describes the generation of new biological species by evolution?

A Selection

B Speciation

C Gene transfer

D Natural selection.

9. Which type of enzyme inhibition involves the binding of the inhibitor to the active site?

A Competitive

B Non-competitive

C Feedback

D End-product.

10. The energy from electrons in the electron transport chain is used to:

A synthesise ATP by combining with phosphate

B activate the enzyme ATP synthase

C combine hydrogen with NAD

D pump hydrogen ions against a concentration gradient.

11. The following diagram shows part of a metabolic pathway.

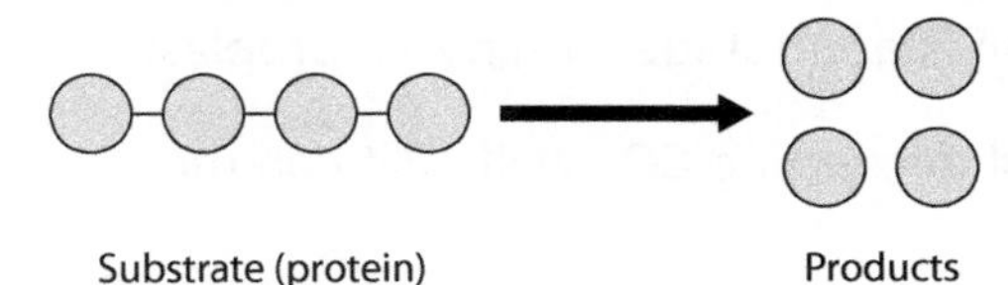

Which line in the table shows the type of reaction shown above and the products?

	Reaction type	***Products***
A	anabolic	amino acids
B	catabolic	amino acids
C	anabolic	glucose molecules
D	catabolic	glucose molecules

12. Which of the following is **not** a suitable method of measuring the rate of metabolism in a human?

A Oxygen used up in a given time

B Heat energy released in unit time

C Rate of carbon dioxide production

D Heart rate in unit time.

13. Which of the following statements about regulators is **true**?

A They do not require energy to achieve homeostasis

B They control their internal environment, which increases the range of possible ecological niches

C Their internal environment is dependent upon their external environment

D They may have low metabolic costs and a narrow ecological niche.

14. During the growth of micro-organisms organisms, in which phase do the cells grow and multiply at the maximum rate?

A Lag

B Exponential

C Stationary

D Death.

15. The diagram below represents part of the carbon fixation stage within a chloroplast.

Which line in the table below shows the effect of decreasing CO_2 availability on the concentrations of RuBP and GP in the cycle?

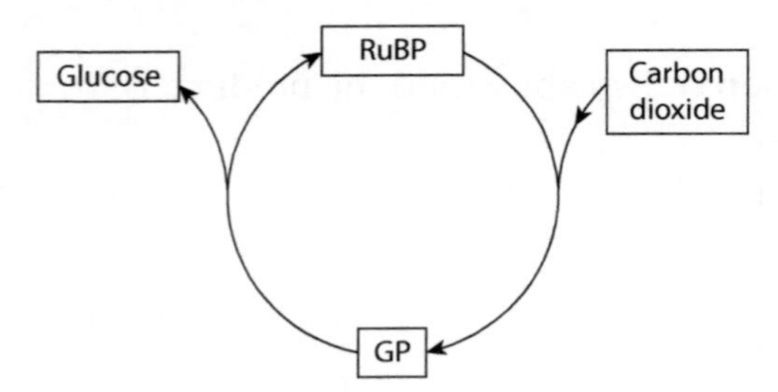

	RuBP concentration	*GP concentration*
A	increase	decrease
B	decrease	increase
C	increase	increase
D	decrease	decrease

16. Which of the following is **not** a factor affecting food security?

A Quantity

B Access

C Quality

D Abiotic.

17. The graph below shows the changes in the populations of red and grey squirrels in an area of woodland over a 10-year period.

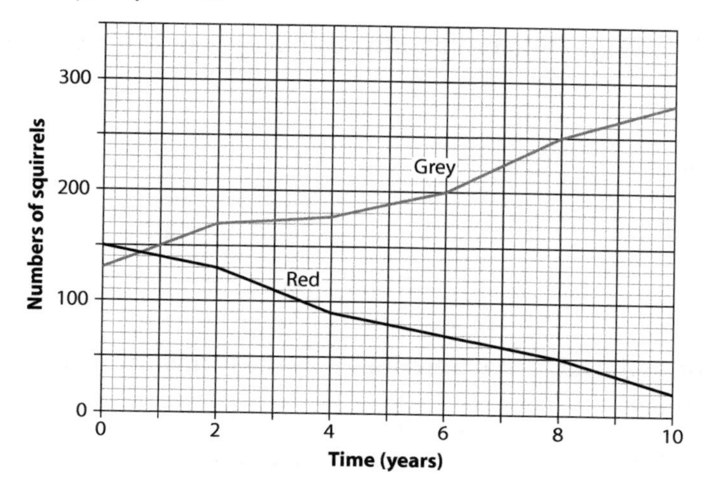

From the graph, the following conclusions were suggested.

1 The grey squirrel population increased by 150% over the 10-year period.

2 The red squirrel numbers decreased from 150% to 20 over the 10-year period.

3 After 8 years, the grey squirrel population was five times greater than the red.

Which of the conclusions are **correct**?

A 1 and 2 only

B 1 and 3 only

C 2 and 3 only

D 1, 2 and 3

18. Inbreeding animals:

A can result in the frequency of homozygotes for recessive harmful alleles

B involves animals that are not closely related

C that are genetically different is likely to lead to inbreeding depression

D leads to an increase in heterozygosity.

19. Which of the following is **not** an ideal characteristic of a pesticide?

A Insoluble

B Specific

C Short-lived

D Safe.

20. Appeasement behaviour in chimpanzees demonstrates they will accept another animal as dominant.

Which of the following would be considered appeasement behaviour in chimpanzees?

A Roaring

B Standing tall

C Chest beating

D Grooming.

21. Identify the line in the table that describes how the recipient and donor are affected by altruistic behaviour.

	Affect on recipient	*Affect on donor*
A	harms	benefits
B	benefits	harms
C	benefits	benefits
D	harms	harms

22. The table below shows the average yield (tonnes per unit area) for two years, 1 and 2, for four crop plants, A, B, C and D.

Crop plant	*Average yield (tonnes per unit area)*	
	Year 1	*Year 2*
A	14.5	29.0
B	4.5	9.0
C	2.7	8.1
D	3.6	4.5

Which crop plant showed a 200% increase from year 1 to year 2?

23. Which term describes a species that humans have moved either intentionally or accidentally to new geographic locations?

A Introduced

B Naturalised

C Invasive

D Native.

24. The graph below shows changes that occur in the masses (kg) of protein, fat and carbohydrate in the body of a hibernating mammal during 7 weeks without food.

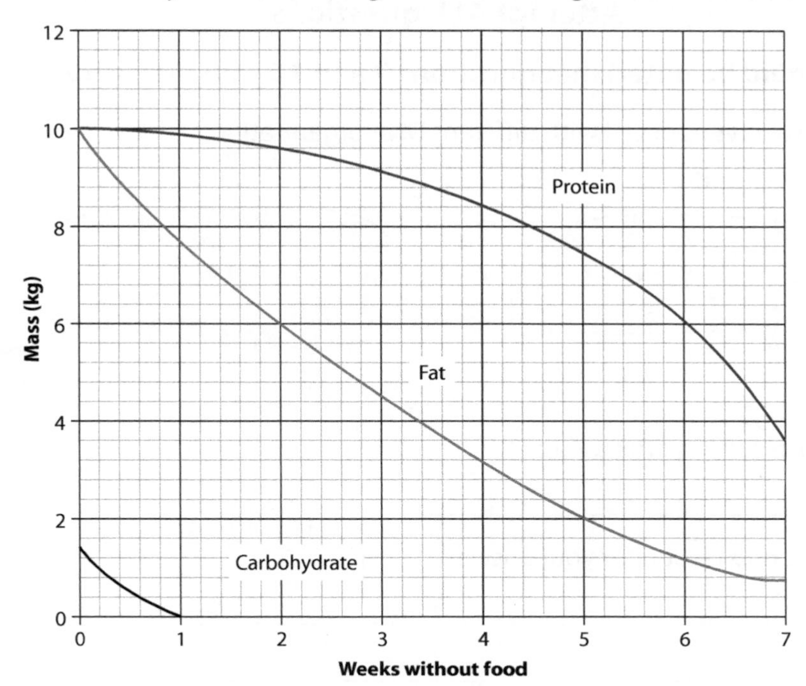

What percentage of the original mass of fat was used up between weeks 2 and 5?

A 33%

B 40%

C 67%

D 80%

25. A control experiment:

A keeps all variables constant except the one being investigated

B usually allows for several variables to be changed at the same time

C does not help eliminate uncertainty about the reliability of the results

D does not allow the cause and effect in an experiment to be tested.

MARKS | DO NOT WRITE IN THIS MARGIN

PAPER 2 – 95 marks

Attempt ALL questions

It should be noted that questions 7 and 16 contain a choice.

1. Below are two types of cell with different DNA organisation.

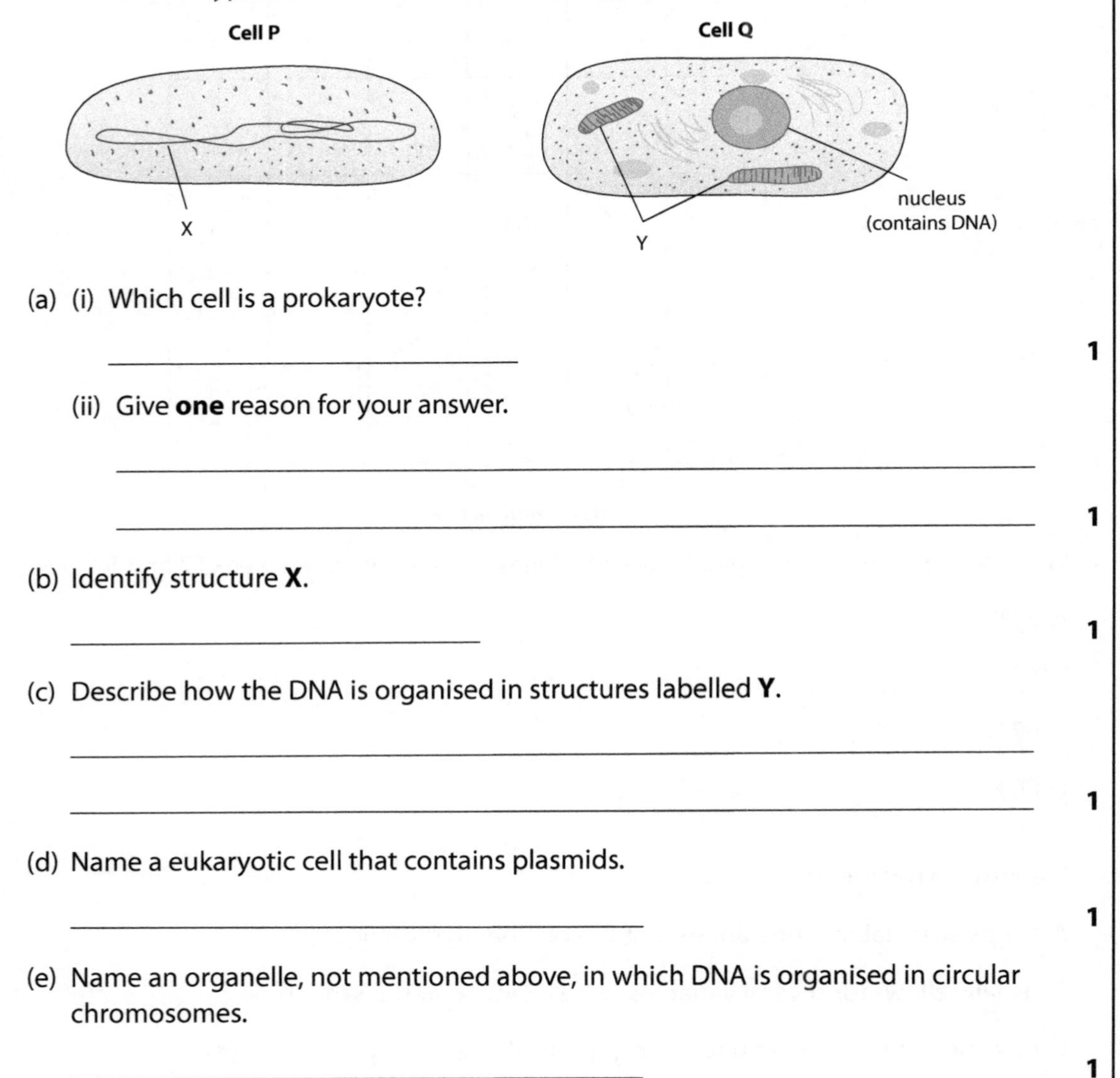

(a) (i) Which cell is a prokaryote? **1**

____

(ii) Give **one** reason for your answer. **1**

____

____

(b) Identify structure **X**. **1**

____

(c) Describe how the DNA is organised in structures labelled **Y**. **1**

____

____

(d) Name a eukaryotic cell that contains plasmids. **1**

____

(e) Name an organelle, not mentioned above, in which DNA is organised in circular chromosomes. **1**

____

MARKS | DO NOT WRITE IN THIS MARGIN

2. The diagram below shows the first stage of gene expression in a cell.

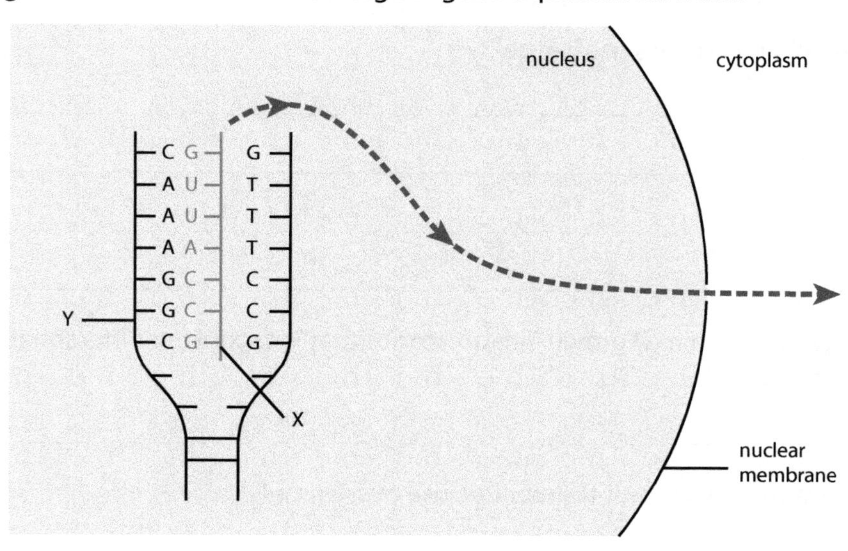

(a) Name this first stage.

____________________ **1**

(b) Name molecule **X**.

____________________ **1**

(c) Describe **two** ways in which molecule **X** is different from molecule **Y**.

1 ____________________

2 ____________________ **2**

(d) Name the parts of molecule **X** that are cut out by splicing.

____________________ **1**

(e) Name the enzyme responsible for the synthesis of molecule **X**.

____________________ **1**

3. Stem cells are unspecialised cells in animals.

(a) State **two** characteristics of stem cells.

1 ______________________________

____

2 ______________________________

______________________________ **2**

(b) Name a type of stem cell grown in laboratories that is capable of differentiating into all types of cell.

______________________________ **1**

(c) Give an example of current therapeutic use of stem cells.

______________ **1**

(d) Give **one** example of an ethical issue that using stem cells raises.

____

______________________________ **1**

4. The diagram below shows a stage in gene expression.

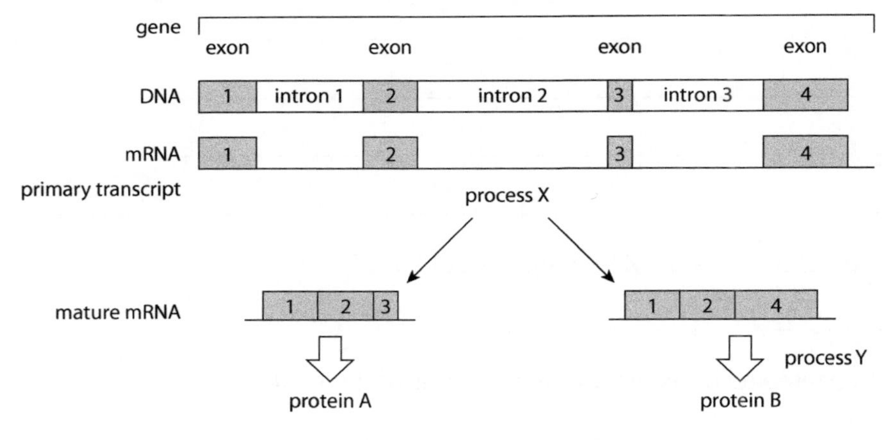

(a) State the cellular location of process X shown above.

______________________________ **1**

(b) Mutations can occur that result in some exons being omitted from the mature transcript. State the name of this type of single gene mutation.

______________________________ **1**

(c) Identify a non-coding region of DNA from the diagram.

______________________________ **1**

MARKS | DO NOT WRITE IN THIS MARGIN

(d) State the name of process X and process Y.

X ____________________

Y ____________________ **2**

5. Evolutionary relatedness among different groups of organisms can be studied and the data used to construct a diagram such as that shown below.

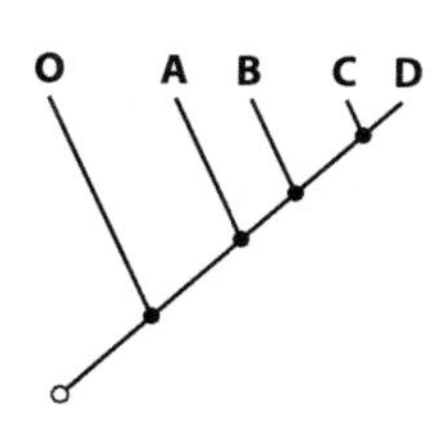

(a) (i) Name this area of study.

____________________ **1**

(ii) From the diagram above, state which **two** organisms are the most closely related.

____________________ and ____________________ **1**

(b) Below is a molecular clock illustrating the relatedness of some mammals.

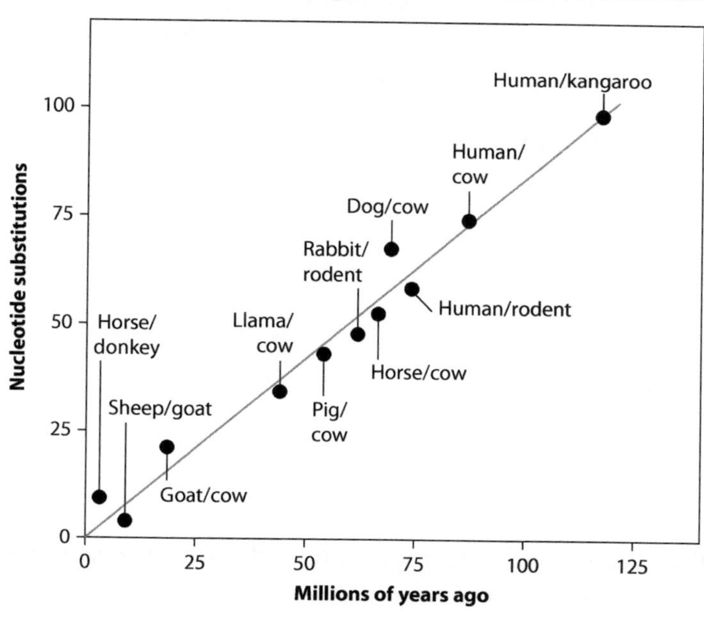

(i) State the purpose of a molecular clock.

____

____________________ **1**

(ii) According to this clock, which **two** pairs of organisms share the most recent common ancestor?

1 ______________________________

2 ______________________________ **2**

6. The diagram below shows the first two stages of cellular respiration.

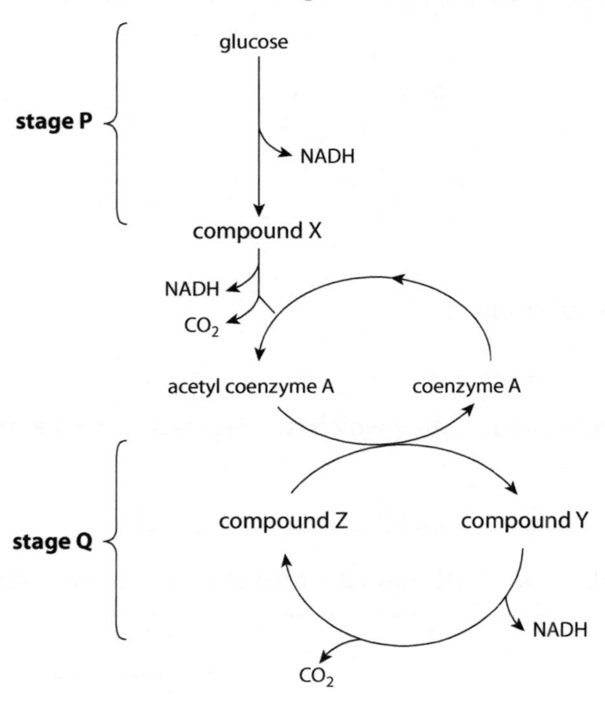

(a) Name compounds **X**, **Y** and **Z**.

X ________________

Y ________________

Z ________________ **3**

(b) Name stages **P** and **Q**.

P ______________________________

Q ______________________________ **2**

(c) What eventually happens to the hydrogen ions and electrons released in the final stage of the electron transport chain?

____ **1**

MARKS | DO NOT WRITE IN THIS MARGIN

7. Answer **either A or B.**

A Give an account of metabolism in conformers and regulators. **4**

OR

B Give an account of carbon fixation. **4**

Labelled diagrams may be used where appropriate.

8. A small farm pond was accidentally polluted with a fertiliser. The changing numbers (numbers/l) of two different species of pond invertebrates – *Daphnia magna* and *Daphnia pulex* – were measured over a period of 9 days following the pollution event. The results are shown in the table below.

	Numbers	
Day	***Daphnia magna* (per l)**	***Daphnia pulex* (per l)**
1	20	10
2	30	15
3	120	30
4	130	55
5	170	60
6	180	65
7	182	70
8	185	90
9	185	120

(a) On the grid below, plot **two** line graphs to show the change in numbers of each species against the number of days after the pollution event. **2**

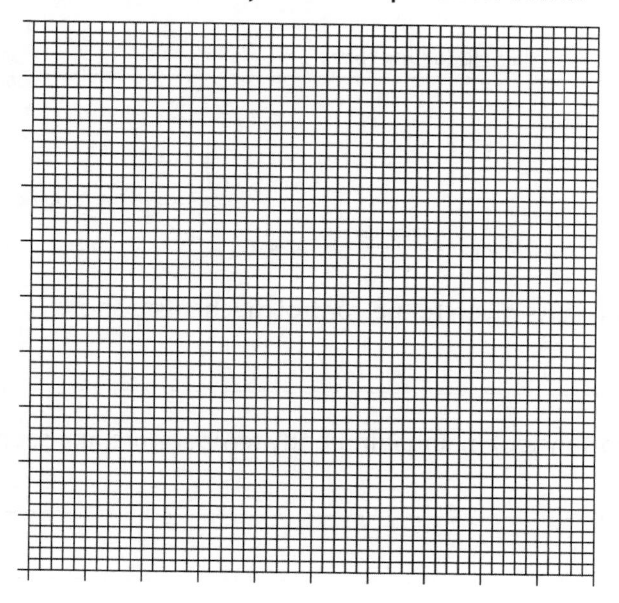

(b) From these results, draw **two** conclusions about the effect of the pollution on the numbers of the two species of *Daphnia*.

1 ____

____

____

____ **1**

2 ____

____

____

____ **1**

(c) Calculate the percentage increase in the numbers of *Daphnia pulex* from day 1 to day 5.

Space for calculation

______________% **1**

9. Organisms can regulate their internal environment using the control mechanism below.

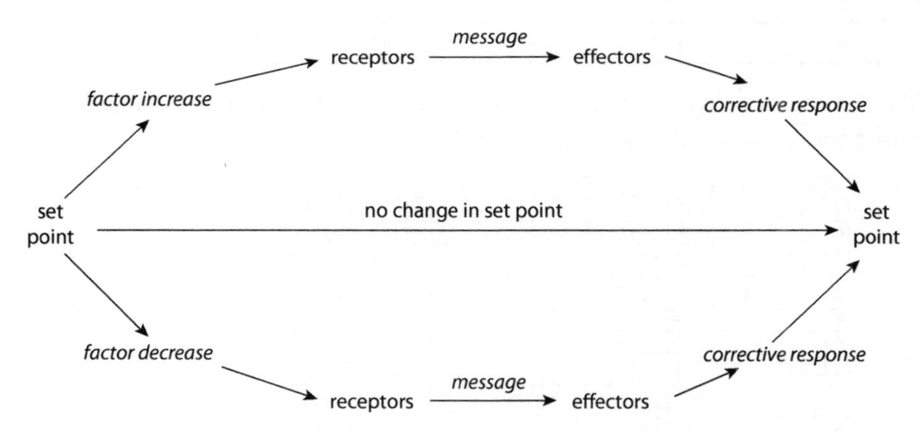

(a) State the general name given to the control mechanism shown.

____ **1**

(b) Explain the importance of maintaining body temperature at a set point

____

____ **2**

(c) When body temperature decreases muscles in parts of the body other than the skin can act as effectors.

(i) Describe the response of skeletal muscles to a decrease in body temperature.

______ 1

(ii) Explain how these effectors bring about a corrective response

______ 1

(d) The hypothalamus is the temperature-monitoring centre in mammals.

Explain how the hypothalamus detects changes in temperature of the blood and how it responds to maintain thermoregulation.

Detects ______

Responds ______

______ 2

10. (a) Decide if each of the statements relating to genetic control of metabolism in the table below is true or false and tick (✓) the appropriate box.

If you decide that the statement is false, write the correct term(s) in the correction box to replace the term underlined in the statement.

Statement	***True***	***False***	***Correction***
Exposure to UV light may result in mutations, some of which may produce an improved strain.			
Recombinant DNA technology uses artificial plasmids as vectors.			
Recombinant yeast cells can be used to produce active forms of a eukaryotic protein that are inactive in bacteria.			

3

(b) The table below shows the results of an experiment where two different yeast cultures are exposed to UV light for 24 hours. Two cultures, A and B, are grown into ten colonies before the start of the experiment.

	Number of colonies	
Exposure time to UV light (hours)	***Yeast culture A***	***Yeast culture B***
0	10	10
6	8	10
12	5	10
18	2	10
24	0	10

(i) State which culture is sensitive to UV light.

_______________ **1**

(ii) Give **two** variables that would have to be kept constant during this experiment.

Variable 1 _______________

Variable 2 _______________ **2**

(iii) State **one** way in which the accuracy of these results could be improved.

____

_______________ **1**

(c) UV light can cause mutations to occur more frequently than normal.

Give an example of another such agent.

_______________ **1**

11. The graph below shows the absorption spectra of three different leaf pigments.

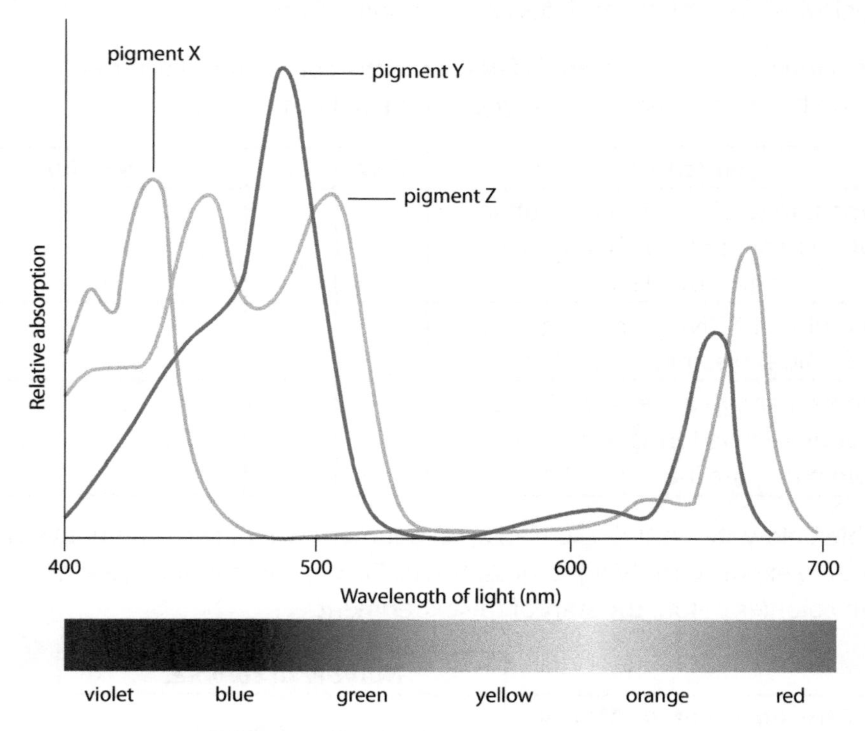

(a) Identify pigments **X**, **Y** and **Z**.

X _______________

Y _______________

Z _______________ **2**

MARKS | DO NOT WRITE IN THIS MARGIN

(b) Some of the light energy coming into contact with a leaf is absorbed.

State one other fate of the light energy that is not absorbed by a leaf.

____ **1**

(c) An experiment was set up to determine the rate of photosynthesis in *Elodea* using different wavelengths of light. The rate of photosynthesis was measured by the number of bubbles produced per minute.

The apparatus is shown below.

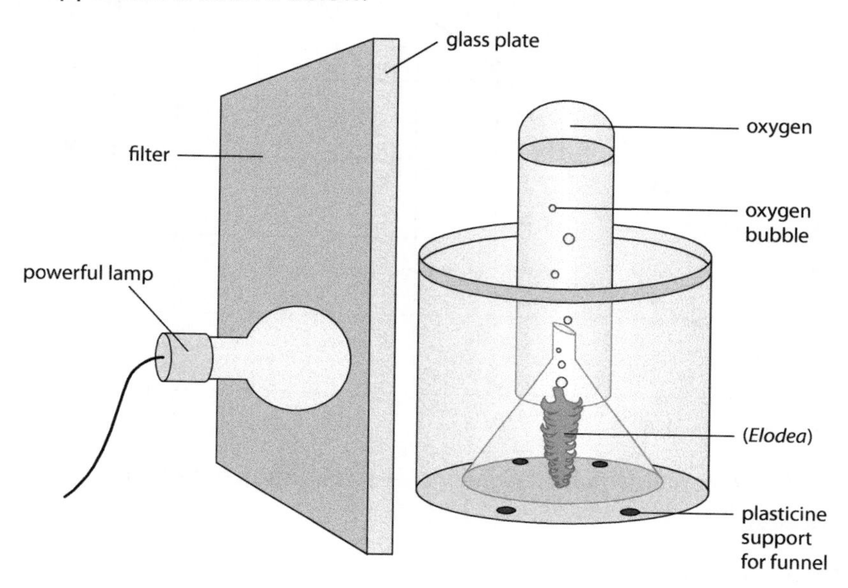

(i) Describe the purpose of the glass plate in this experiment.

____

____ **1**

(ii) Filters absorb light of different wavelengths (colours) and reflect other wavelengths.

State the colour of filter that would produce the greatest rate of photosynthesis on an action spectrum produced using results from this experiment.

_______________________ **1**

(iii) Identify the dependent variable in this experiment.

____ **1**

12. The graph below shows how world population and fertiliser usage changed between 1920 and 2000.

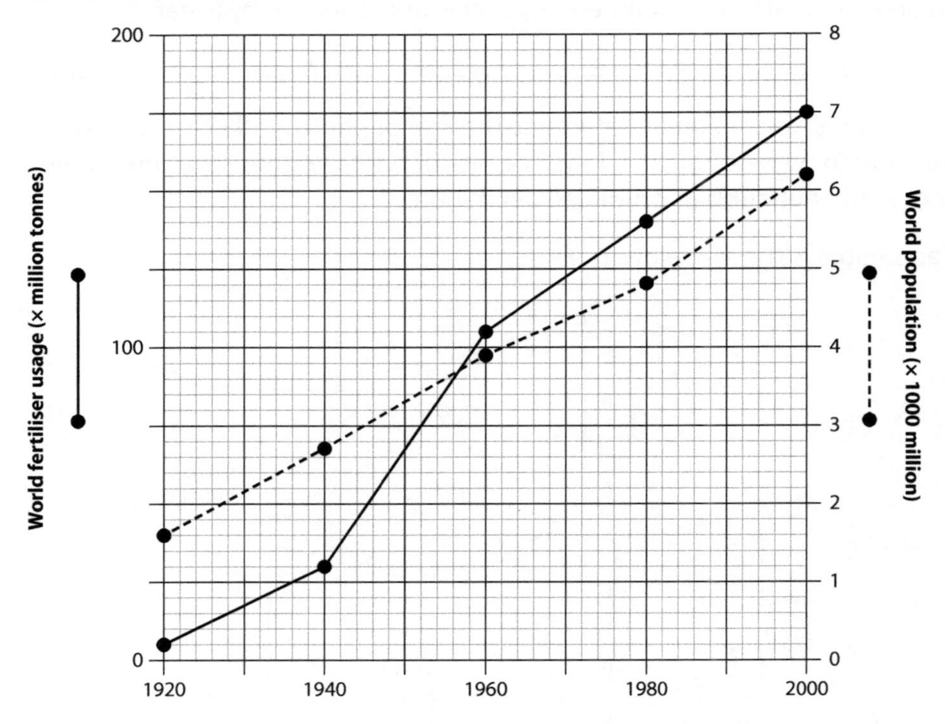

(a) (i) Calculate the percentage increase in world fertiliser usage between 1960 and 2000.

Space for calculation

______________% **1**

(ii) State the world population when the world fertiliser usage was 100 million tonnes.

______________ (× 1000 million) **1**

(b) State **one** conclusion that can be drawn from the graph above.

____

____________________________________ **1**

(c) State **two** other chemicals, other than fertilisers, used in agriculture to increase crop yield.

1 ________________________

2 ________________________ **1**

(d) State **two** methods used in integrated pest management.

____

____________________________________ **1**

13. In an investigation, the behaviours of 10 monkeys in a small zoo enclosure were observed over 24 hours.

(a) Some of the observed monkeys showed very high levels of activity and others showed very low levels of activity.

State what such altered levels of activity might indicate.

______________________ **1**

(b) Some monkeys were observed chewing their own tail.

What term is used to describe this type of behaviour?

______________________ **1**

(c) Identify a feature of this investigation which suggests that results were reliable.

____

____ **1**

(d) Give **two** examples of behaviours that reduce conflict between monkeys that may have been observed in this investigation.

1 ____ **1**

2 ____ **1**

(e) One of the smallest monkeys developed an alliance with the dominant monkey.

State **one** advantage of this alliance.

____

____ **1**

14. An investigation into the metabolic rate of a stick insect at rest, using oxygen uptake (cm^3) was carried out using the apparatus shown.

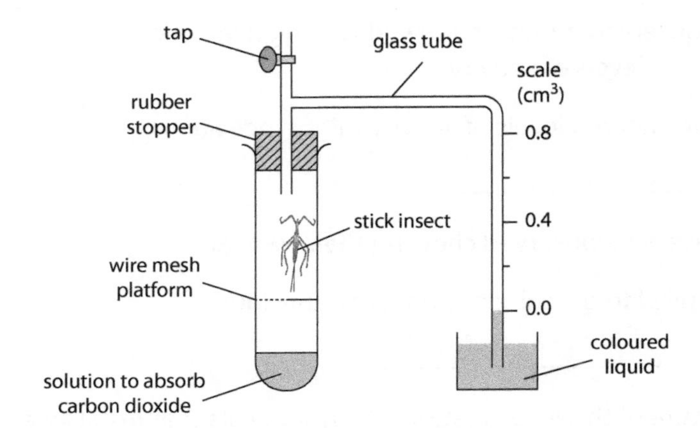

To start the experiment, the tap was closed and the reading on the scale recorded.

The apparatus was kept at 15°C with the tap open for 10 minutes before the investigation was started.

Every 2 minutes, for 10 minutes, readings from the scale were recorded.

The results are shown in the table below.

***Time after tap closed* (minutes)**	***Oxygen uptake* (cm^3)**
0	0.00
2	0.08
4	0.16
6	0.24
8	0.36
10	0.40

(a) Identify the dependent variable in this experiment.

____________________ **1**

(b) Describe how the experimental procedure could be improved to increase the reliability of the results.

____________________ **1**

(c) An identical apparatus was set up without the stick insect as a control.

Explain why the use of a control ensures valid results.

____

____________________ **1**

(d) Explain why the apparatus was left for 10 minutes with the tap open before readings were taken.

____

____ 1

(e) On the grid below, plot a line graph to show the oxygen uptake against time.

2

(f) From the results of this investigation, describe the changes in oxygen uptake of the stick insect over the period 0–8 minutes.

____

____

____ 2

(g) The mass of the stick insect was 10.0 g.

Use the results in the table to calculate the average rate of oxygen uptake per gram of stick insect per minute over the 10 minute period.

Space for calculation

______________ cm^3 per gram per minute 1

(h) Predict the effect of a decrease in temperature from 15°C to 5°C on the oxygen uptake by the stick insect and justify your answer.

Prediction ____

Justification ____

____ 2

15. The graph below shows the annual consumption of fertiliser (kg × 10^9 / year) in major regions of the world for three different years.

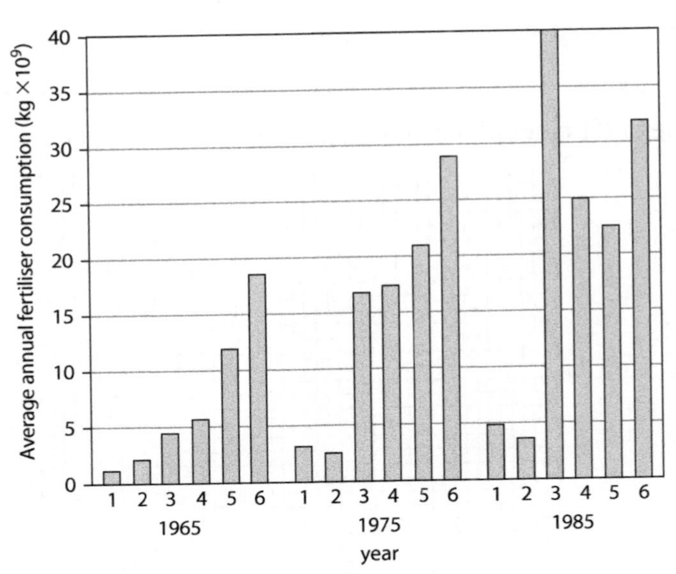

(a) Using values from the graph, describe the changes in the use of fertiliser in Africa between 1965 and 1985 in comparison to Asia. **2**

(b) Calculate the yearly increase in average annual fertiliser consumption in Europe between 1965 and 1985.

Space for calculation

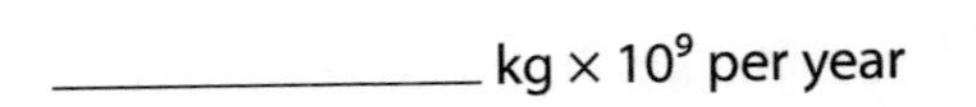
 1

MARKS | DO NOT WRITE IN THIS MARGIN

16. Answer **either A or B**.

A Describe symbiosis under the following headings. **9**

(i) Parasitism

(ii) Mutualism

OR

B Describe biodiversity under the following headings. **9**

(i) Components of biodiversity

(ii) Threats to biodiversity

Labelled diagrams may be used where appropriate.

[END OF QUESTION PAPER]

PAPER 1 ANSWER GRID

Mark the correct answer as shown ◉

	A	B	C	D
1	◯	◯	◯	◯
2	◯	◯	◯	◯
3	◯	◯	◯	◯
4	◯	◯	◯	◯
5	◯	◯	◯	◯
6	◯	◯	◯	◯
7	◯	◯	◯	◯
8	◯	◯	◯	◯
9	◯	◯	◯	◯
10	◯	◯	◯	◯
11	◯	◯	◯	◯
12	◯	◯	◯	◯
13	◯	◯	◯	◯
14	◯	◯	◯	◯
15	◯	◯	◯	◯
16	◯	◯	◯	◯
17	◯	◯	◯	◯
18	◯	◯	◯	◯
19	◯	◯	◯	◯
20	◯	◯	◯	◯
21	◯	◯	◯	◯
22	◯	◯	◯	◯
23	◯	◯	◯	◯
24	◯	◯	◯	◯
25	◯	◯	◯	◯

Higher Biology

Practice Papers for SQA Exams

Exam B

Fill in these boxes and read what is printed below.

Full name of centre	Town

Forename(s)	Surname

Answer all of the questions in the time allowed.

Total marks – 120

Section 1 – 25 marks – Duration – 40 minutes

Section 2 – 95 marks – Duration – 2 hours 20 minutes

Read all questions carefully before attempting.

Write your answers in the spaces provided, including all of your working.

PAPER 1 – 25 marks

Attempt ALL questions

Answers should be given on the separate answer sheet provided.

1. Which line in the table correctly shows where deoxyribose and phosphate are found in each strand of DNA?

	3′ end	*5′ end*
A	deoxyribose	phosphate
B	phosphate	deoxyribose
C	deoxyribose and phosphate	neither present
D	neither present	deoxyribose and phosphate

2. DNA polymerase

A causes the formation of hydrogen bonds between bases

B cannot add nucleotides to an already existing DNA template strand

C requires a primer to be present

D can only add nucleotides to the free 5′ end of the new DNA strand.

3. A student investigated the effect on growth of different volumes of water added to three groups of cloned tobacco plants over a 1-week period. Each group contained the same number of plants. The first group received 300 cm^3 water; the second group received 200 cm^3 water and the third group received 100 cm^3 water.

What would be a suitable control for this experiment?

A Increase the size of all the groups

B Use tobacco plants that were not clones

C Use a fourth group of plants to which 50 cm^3 water was added

D Use a fourth group of plants that had no water added.

4. Gene expression is controlled by regulating

A translation only

B transcription only

C both translation and transcription

D post-translational modification.

5. The data below shows how long (minutes) muscles can work without getting tired (endurance) as a function of the initial concentration (g/kg muscle) of liver glycogen (an energy source for respiring muscles).

Endurance (mins)	60	80	100	120	140	160
Initial concentration of liver glycogen (g/kg muscle)	5	10	15	20	25	30

Predict the likely initial concentration (g/kg muscle) in an individual who has an endurance time of 240 minutes.

A 40

B 45

C 50

D 55

6. A double-stranded DNA fragment was found to have 180 nucleotides present. Which combination of bases would fit into this fragment?

A 40 thymine and 50 adenine

B 50 adenine and 50 cytosine

C 40 cytosine and 50 guanine

D 50 cytosine and 40 adenine.

7. Introns are

A sections of DNA that code for mRNA

B enzymes that join up DNA fragments

C non-coding lengths of DNA

D segments of DNA that repeat themselves.

8. Starting with the earliest, arrange the following events in the evolutionary order in which they appeared.

1. Photosynthetic organisms

2. Eukaryotes

3. Vertebrates

4. Animals

A 2 1 3 4

B 2 3 1 4

C 1 2 4 3

D 1 2 3 4

9. Which of the following is a correct description of a selection which can alter the frequency of a genotype?

A Stabilising selects against an average phenotype

B Directional selects one extreme of the phenotype

C Disruptive two or more phenotypes are selected against

D Stabilising selects for extremes of the phenotype.

10. An example of an anabolic reaction is:

A synthesis of the enzyme pepsin

B breakdown of glucose in respiration

C digestion of protein to form amino acids

D conversion of ATP to ADP.

11. During aerobic respiration, NADH is not produced in

A glycolysis

B the electron transport chain

C the citric acid cycle

D the conversion of pyruvate to acetyl coenzyme A.

12. A human nerve cell was magnified 80,000 times and measured 260 mm.

The length of the cell in micrometres is

A 1.25

B 1.50

C 3.00

D 3.25

13. Which line in the table below correctly describes features of the circulatory system found in a mammal?

	Features	
	Heart	***Circulation***
A	four chambers	incomplete double
B	two chambers	complete double
C	four chambers	incomplete single
D	two chambers	complete double

14. Conformers

A are able to regulate their metabolism and maintain a steady internal environment

B use physiological mechanisms to alter their metabolic state

C depend on behaviour to regulate their internal state

D live in unstable environments.

15. Which of the following would be most likely to contribute to increasing food security in an environmentally friendly way?

A Transferring genetic material from domestic strains into wild strains

B Extending the use of insecticides

C Using more organically based fertilisers

D Decreased plant productivity.

16. A farmer loses the following percentage yields to his crop due to three different agents: insects, infection and weeds:

Agent	*Percentage loss in yield*
insects	16
infection	32
weeds	24

Which is the correct smallest whole number ratio of the percentage loss due to insects, weeds and infection?

A 4 : 3 : 2

B 2 : 3 : 4

C 2 : 4 : 3

D 4 : 2 : 3

17. Which of the following statements related to inbreeding is correct?

A Involves the fusion of two gametes that are genetically very different

B Involves the fusion of two gametes from different species

C Causes a decrease in the frequency of heterozygosity

D Usually results in a decrease in homozygotes who are recessive for harmful alleles.

18. Which of the following characteristics do perennial weeds usually possess?

A Fast-growing and long life cycles

B Slow-growing and produce high seed output

C Possess storage organs and can reproduce asexually

D Produce few seeds and cannot reproduce asexually.

19. Animal welfare ensures animals

A that are used in the human food chain have a better quality of life

B do not exhibit natural behaviour patterns

C are kept in safe conditions for part of the year

D are reared in isolation to prevent infection.

20. Identify the line in the table that incorrectly describes the type of farming.

	Intensive	***Free range***
A	More labour intensive	Less labour intensive
B	Less land	More land
C	Higher profits	Lower profits
D	Costs low	Costs high

21. Which of the following feeding relationships is classed as mutualism?

A Lice on human hair

B Cellulose-digesting bacteria in the gut of cattle

C Leech sucking blood from fish

D Tapeworm living inside dog.

22. The following pie-chart shows causes of death in a Scottish city for males in 2012.

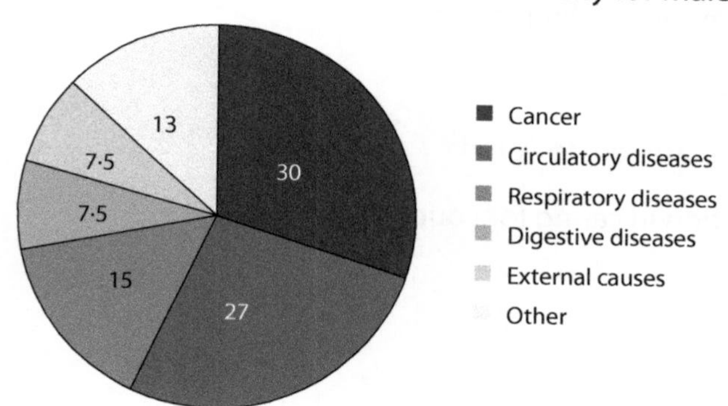

Which of the following statements is true, based on these data?

A Cancer causes three times as many deaths as respiratory and digestive diseases combined

B Deaths due to other causes number approximately half as many as those caused by circulatory diseases

C 40% of deaths are due to cancer and respiratory diseases combined

D External causes are responsible for twice as many deaths compared to digestive diseases.

23. Which of the following would be most likely to have a negative effect on biodiversity?

A Planting new trees along a motorway

B Leaving areas of land unmanaged in a woodland

C Connecting isolated habitats that were previously in contact

D Introducing an invasive species into an environment.

24. Which of the following forms of biodiversity are measurable?

A Genetic, species, ecosystem

B Genetic, species, competition

C Species, ecosystem, competition

D Competition, genetic, ecosystem

25. An advantage to an animal living in a social group is

A Increased competition for resources

B Less energy is wasted

C Reproduction rates are lowered

D Less cooperation in caring for young.

PAPER 2 – 95 marks

Attempt ALL questions

It should be noted that questions 5 and 19 contain a choice.

1. An early experiment that helped establish that DNA could be transferred from a dead cell to a live cell involved the use of a bacterium. This bacterium exists in two forms: smooth (S) and rough (R). One, 'strain S', produces pneumonia in mice, causing death; and the other, 'strain R', is harmless to mice. This bacterium can be killed by heat.

The diagram below shows how the experiment was carried out.

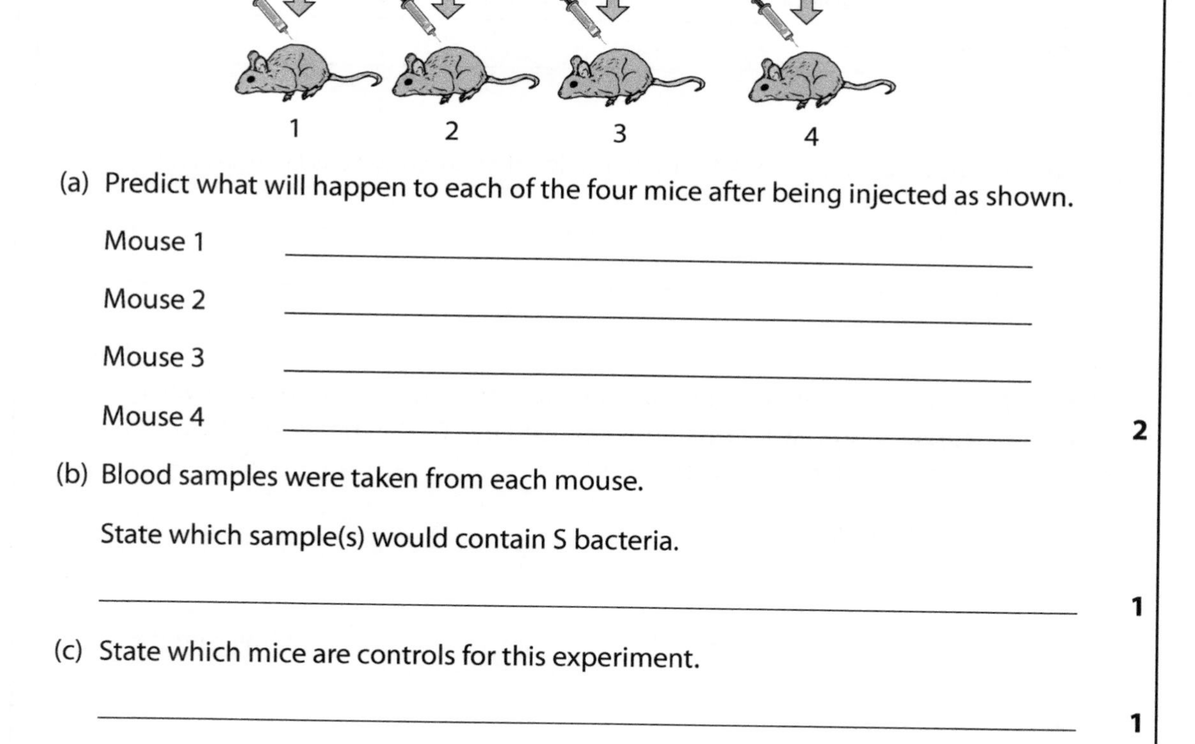

(a) Predict what will happen to each of the four mice after being injected as shown.

Mouse 1 ____________________

Mouse 2 ____________________

Mouse 3 ____________________

Mouse 4 ____________________ **2**

(b) Blood samples were taken from each mouse.

State which sample(s) would contain S bacteria.

______________________________ **1**

(c) State which mice are controls for this experiment.

______________________________ **1**

2. (a) A sequence of bases of an mRNA molecule reads as follows:

AUAAUACUUCAAGAGCUUGAGGAGGAGCUU

There are 4 different amino acids in the protein coded for by this section of mRNA:

isoleucine (Ile), leucine (Leu), glutamic acid (Glu) and glutamine (Gln).

Only one glutamine is present.

Complete the following diagram to show the amino acid sequence using the abbreviations given.

Ile				Glu					

2

(b) Give the tRNA anticodons for the following sequence of amino acids:

Amino acids	Isoleucine	Glutamic acid	Leucine
tRNA anticodon			

1

(c) Describe how the linear DNA found in nuclei of eukaryotic cells is organised.

____

____ **2**

(d) A chromosome in an animal cell was estimated to be 40 mm long. However, in the nucleus, its length was apparently reduced by 200,000 times. What would this apparent length be in micrometres?

Space for calculation

____________ micrometres **1**

MARKS | DO NOT WRITE IN THIS MARGIN

(e) The mass of a cell's genome is measured in picograms (pg).

1 picogram = 10^{-12} gram

To convert the mass of the genome to a number of base pairs, a formula is used, as follows:

$$\text{number of base pairs} = \text{mass (pg)} \times 0.978 \times 10^9$$

Calculate the mass (pg) of a chromosome that has 58,680,000 base pairs.

Space for calculation

________ pg **1**

3. The polymerase chain reaction (PCR) can be used to produce many copies of a piece of DNA in the laboratory. The flow chart below shows how a sample of DNA was treated during one cycle.

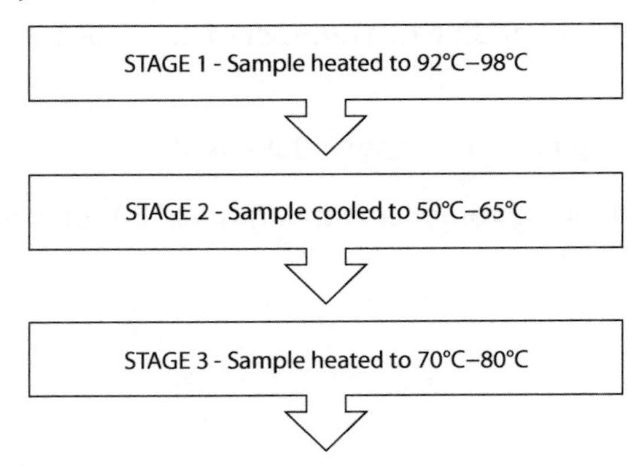

(a) Describe the effect of the different heat treatments in Stage 1 and Stage 2.

Stage 1 ______________________________ **1**

Stage 2 ______________________________ **1**

(b) (i) State the name of the enzyme that causes the DNA to replicate in the PCR process.

______________________________ **1**

(ii) State **one** important property of this enzyme that enables it to be used in this process.

______________________________ **1**

(c) The number of DNA molecules doubles after each cycle of the PCR process.

Calculate how many DNA molecules would be made after three cycles.

Space for calculation

______________ **1**

(d) The tube below shows the mixture of chemicals used in the PCR process.

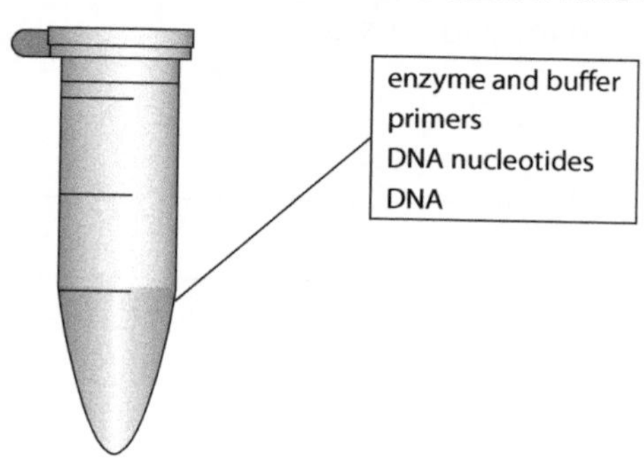

Describe the contents of a control tube designed to show the need for enzyme to be present?

____

____ **1**

4. Gel electrophoresis is a technique used to compare DNA sequences, which can be used by police at a crime scene.

Three DNA bandings are shown below taken from a crime scene and from two possible suspects, A and B.

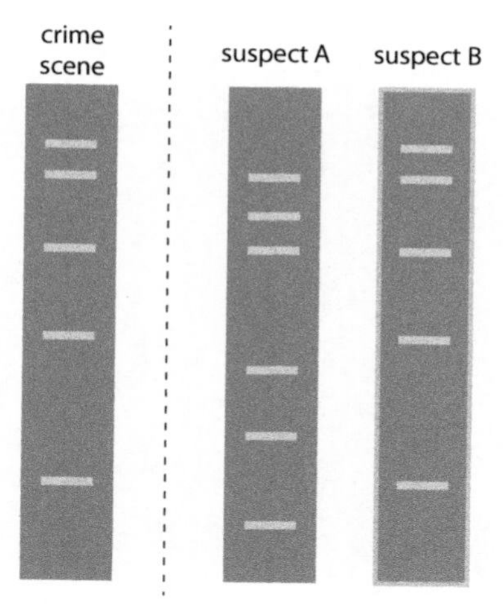

(a) State the general name for large molecules, such as DNA and proteins, that can be separated by gel electrophoresis.

____ **1**

(b) State which of the suspects is most likely to have been present at the crime scene and give a reason for your answer.

Suspect: ____________ 1

Reason: ____

____ 1

5. Answer **either A or B**.

A Describe the structure of RNA. 4

OR

B Describe how the heart chambers in reptiles and birds are arranged and relate this to their metabolic rate. 4

MARKS | DO NOT WRITE IN THIS MARGIN

6. (a) The diagram below shows a part of a root of a geranium plant.

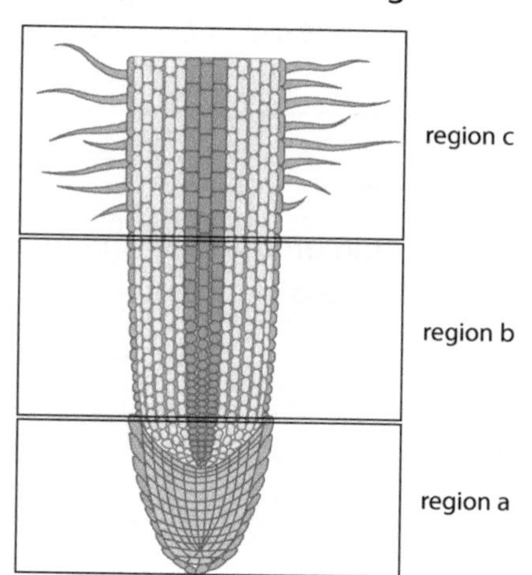

(i) Identify the region that contains the meristem. Insert the letter in the space provided.

Region ____________________ **1**

(ii) State **two** characteristics of the cells that make up this meristem.

1 ____________________

____

2 ____________________

____________________ **2**

(b) Cells in the meristem can selectively switch off genes and become root hair cells.

Suggest **two** benefits to the cell of not having all genes switched on all of the time.

____

____________________ **2**

(c) The table below shows the average number of shoots of a geranium plant produced per meristem over five weeks in two different types of growth media A and B.

Time (weeks)	***Average number of shoots produced per meristem***	
	Growth medium A	***Growth medium B***
1	3	4
2	6	5
3	8	8
4	7	9
5	6	10

Use values from the table to describe the average number of shoots produced per meristem in medium A over the five-week period. **2**

____

____

(d) Predict which growth medium would produce geranium plants with the greatest number of shoots after 6 weeks' growth.

Give a reason for your answer.

Growth medium ________________ **1**

Reason ________________________________

____ **1**

7. The graph below shows the results of an experiment into how changing the light intensity and concentration of carbon dioxide affects the rate of photosynthesis in a plant.

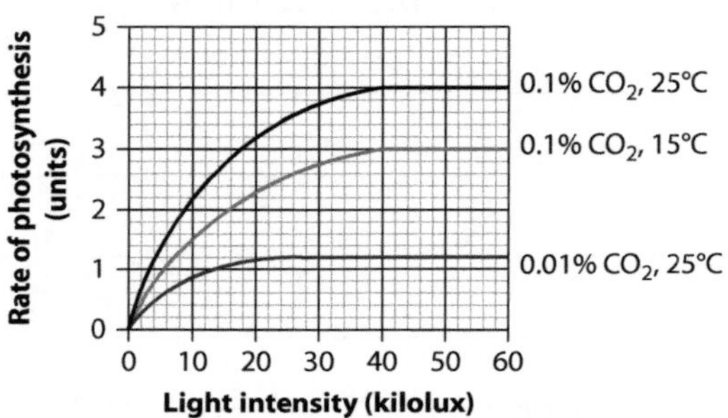

(a) At a light intensity of 50 kilolux, which factor, as shown in the graph, has the greater effect in increasing the rate of photosynthesis?

Justify your answer.

Factor ________________________________

Justification ____________________________

____

____ **1**

(b) Identify the combination of light intensity, carbon dioxide concentration and temperature that gives the maximum rate of photosynthesis.

____ **1**

(c) The experiment was repeated at a carbon dioxide concentration of 0.01% and a temperature of 20°C. Draw, **on the grid above**, a curve to show what the predicted results would be. **1**

(d) The results shown were based on one run of the experiment. Describe how the reliability of the results can be improved.

____

____ **1**

8. The phylogenetic tree below shows some of the evolutionary relationships between different animals.

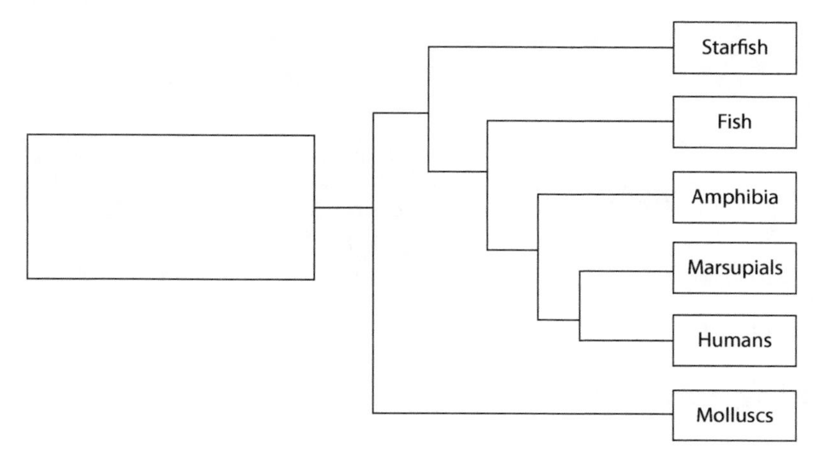

(a) The tree shows animals that share a common lineage and have evolved from an earlier form. Insert the general term for the earlier form in the box provided above. **1**

(b) Humans are more closely related to amphibia than to starfish.

Using the information on the tree, explain how this is known.

____

____ **1**

(c) Identify the group of animals to which marsupials are most closely related.

Use information from the phylogenetic tree to explain your answer.

Group of animals ________________________________

Explanation ____________________________________

____ **1**

9. (a) The following diagram shows two different types of reaction that can occur in cells.

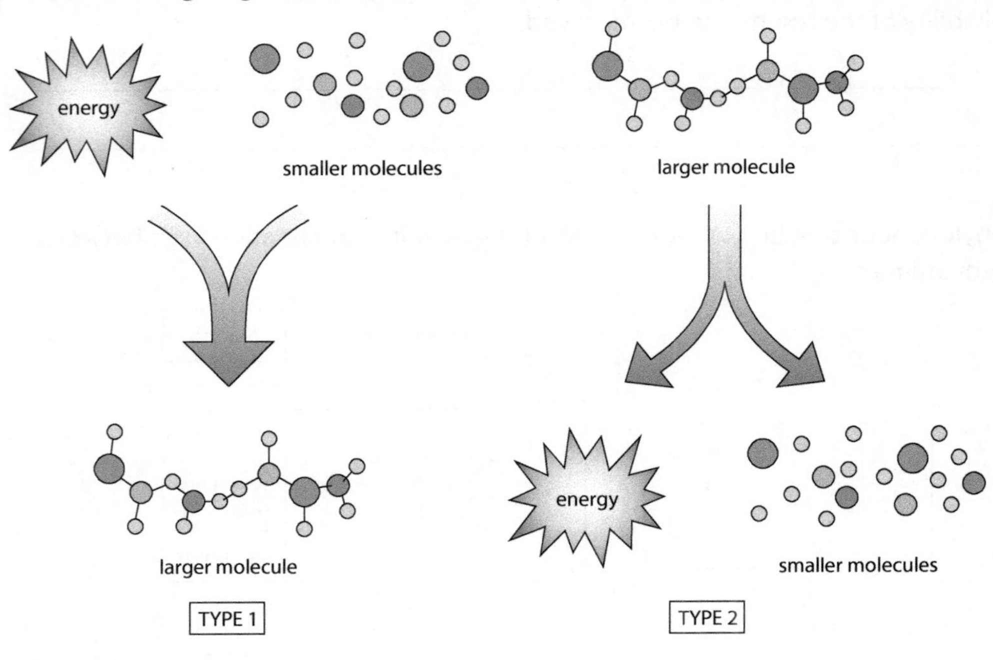

Identify each type of reaction.

Type 1: ____________________

Type 2: ____________________ **2**

(b) Enzymes allow reactions to take place that otherwise might not occur, or would take a long time.

Describe **one** way in which enzymes make it easier for reactions to take place.

____

____________________ **1**

(c) When an enzyme encounters a suitable substrate, the active site changes shape to fit the substrate molecule.

State the term used to describe this function?

____________________ **1**

MARKS | DO NOT WRITE IN THIS MARGIN

(d) When an end-product of an enzyme-catalysed reaction accumulates, it may cause the activity of the enzyme involved to decrease or stop temporarily.

State the term used to describe this type of control?

______________________________ **1**

(e) Some enzyme control is brought about by the action of inhibitor molecules that have a shape very similar to the enzyme's normal substrate and so can bind to the active site.

State the type of inhibition this is.

______________________________ **1**

10. The graph below shows the average number of units of alcohol consumed per week by men and women in 2006 for an area in the UK.

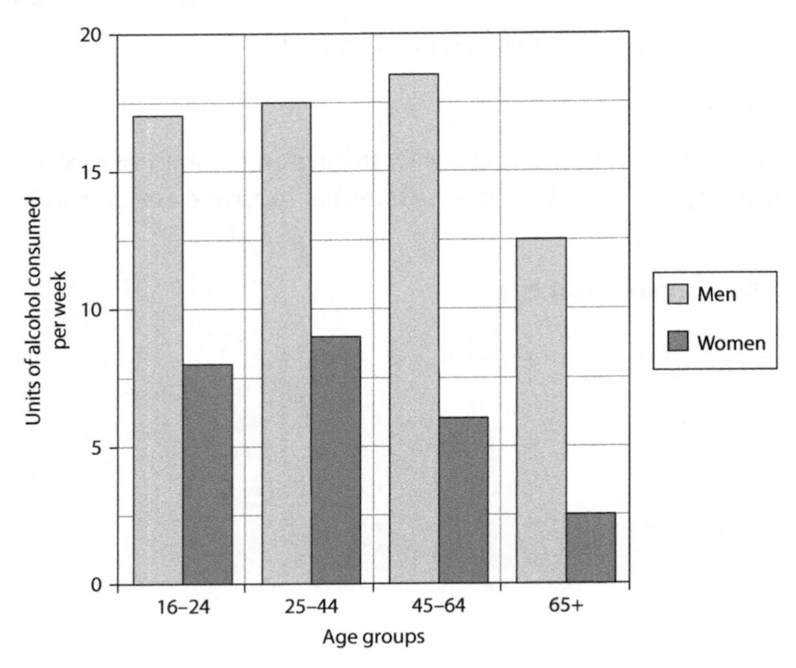

(a) Using information from the graph, calculate the percentage decrease in the number of units of alcohol consumed by a man aged 28 years and a man aged 66 years.

Space for calculation

____________ % **1**

(b) How many times fewer is the number of units of alcohol consumed by a woman aged 66 years compared with a man of a similar age?

Space for calculation

____________ **1**

(c) From the data in the graph, state two different conclusions about the relationship between the age and sex of a person and the average number of units of alcohol consumed in this part of the UK in 2006.

Conclusion 1 ____________

____

____

Conclusion 2 ____________

____

____________ **2**

MARKS | DO NOT WRITE IN THIS MARGIN

11. (a) Birds are able to control their internal environment through their metabolism.

(i) State the term used to describe an animal that can control its internal environment through metabolism?

________________________________ **1**

(ii) State **one** reason why is it important for birds to control their body temperature.

____

____

________________________________ **1**

(b) State the exact location of the temperature-monitoring centre in mammals.

________________________________ **1**

12. The diagram below shows the appearance of a plate containing a medium that enhances the growth of bacteria that can fix nitrogen gas from the air into nitrogen compounds in their cells. This medium had all the essential nutrients for bacterial growth but no nitrogen-containing ammonium compounds. The plate is shown after 3, 7 and 12 days of growth.

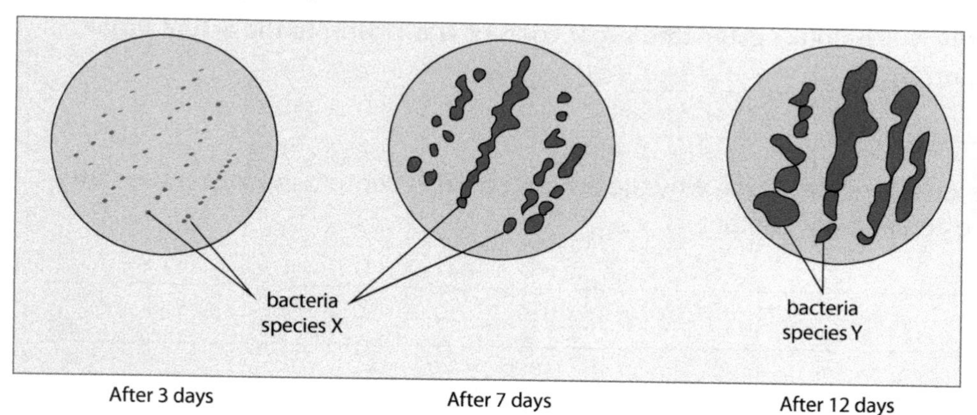

(a) After 3 days only, species X was seen to grow. Suggest why this happened.

____

____

________________________________ **2**

(b) After 12 days, species Y appears. Give a possible explanation for this.

____

____

________________________________ **2**

13. An experiment was carried out to estimate the sugar content, measured in g/L, of various drinks. The graph below shows the results of the experiment along with the actual sugar content.

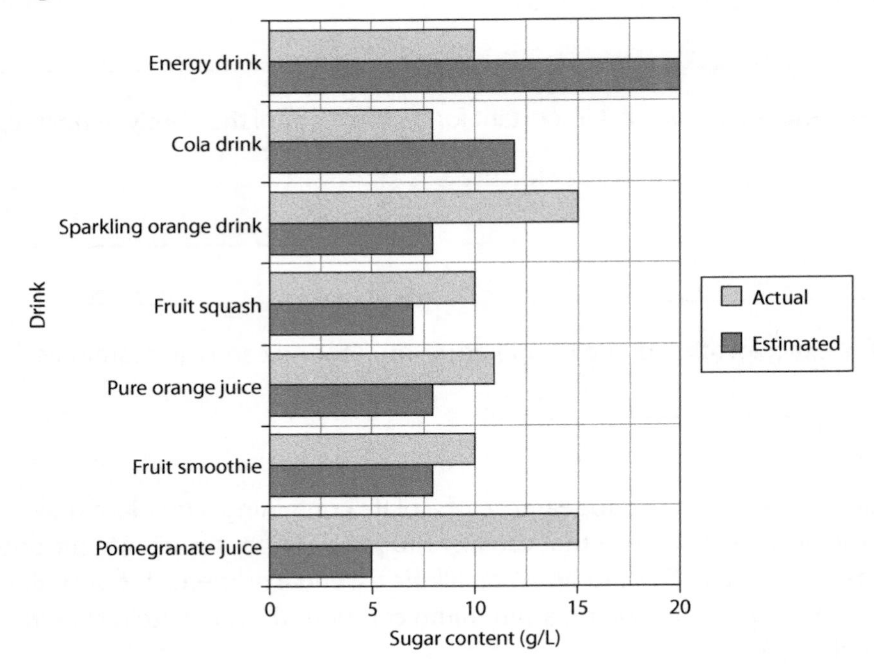

(a) Identify which drink's estimated sugar content was closest to the actual sugar content?

________________________ **1**

(b) State **one** possible reason why the estimated sugar content was never the same as the actual sugar content.

____

____ **1**

14. (a) The grid below shows some of the adaptations organisms have in order to survive adverse conditions.

A small ear flap	B hibernation	C thick layer of skin fat
D aestivation	E migration	F large body size

Using the **letters**, place each of these adaptations under the correct descriptive heading in the table below.

Structure	***Behaviour***

1

(b) State **one** technique used by scientists to monitor long-distance migration in birds and explain how the technique is used.

Technique ______________________

Explanation ______________________

____

______________________ **2**

(c) Migration is affected by both **innate** and **learned** influences.

Give **one** example of each influence.

Innate ______________________

____

Learned ______________________

______________________ **2**

15. (a) The diagram below shows some of the reactions of photosynthesis that take place in a chloroplast.

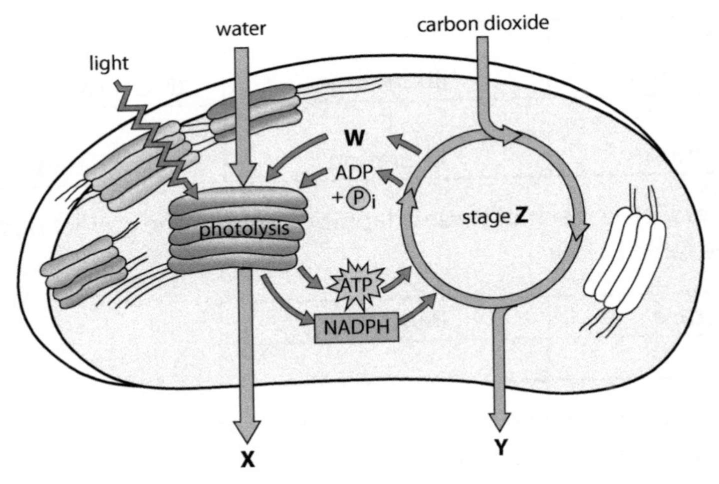

(i) Name molecules W and X.

W ______________________

X ______________________ **2**

(ii) Name stage Z.

______________________ **1**

(iii) Molecule Y can be converted into an important chemical found in the cell wall.

Name this chemical.

______________________ **1**

(b) In eukaryotic cells, molecules involved in speeding up the rate of photosynthesis photosynthesis are contained within the chloroplast.

State the location in the chloroplast where the genetic code for these molecules is located..

______________________ **1**

(c) A plant was allowed to photosynthesise in normal light. The light was then switched off. There was a rise in the concentration of glyceraldehyde-3-phosphate (G3P).

Give **two** reasons why the concentration of G3P decreased.

1 ______________________

____

2 ______________________

______________________ **2**

16. The following are some of the stages in the production of insulin by genetically engineered bacteria:

1. bacteria take in modified plasmids
2. plasmids opened using an enzyme
3. insulin-producing gene isolated from human chromosome
4. gene for insulin production inserted into the plasmid using an enzyme.

(a) State the correct order of these stages, starting with the earliest.

____ **1**

(b) State the enzymes involved in stages 2 and 4.

Stage 2: ________________

Stage 4: ________________ **2**

(c) The bacteria used had a gene that conferred resistance to an antibiotic called ampicillin.

Explain how this could be used to select for those bacteria that had successfully taken up the modified plasmids.

____

____

____ **2**

(d) Suggest **two** advantages of using genetic engineering to produce insulin.

1 ____

2 ____ **2**

(e) Explain the function of a restriction site on the plasmids used for insulin production.

____

____ **1**

17. Growers often face problems with organisms such as weeds, pests and diseases. To combat these, the grower may use a chemical to protect a valuable crop.

(a) State the general term used to describe such a chemical?

______________________________ **1**

(b) Some of these chemicals can be persistent.

Explain what the term 'persistent' means.

____

______________________________ **1**

(c) Give **two** other possible negative effects of using such chemicals.

1 ______________________________

____

2 ______________________________

______________________________ **1**

(d) Continued use of such chemicals can lead to an increase in selection pressure.

State **one** possible outcome of this selection pressure.

____

____

______________________________ **1**

MARKS | DO NOT WRITE IN THIS MARGIN

18. In an investigation, the rate of photosynthesis by plant leaf discs was measured at different light intensities. The results are shown in the table.

Light intensity **(kilolux)**	***Rate of photosynthesis by leaf discs*** **(units)**
5	1
10	20
15	25
20	32
25	45
30	45

(a) Plot a line graph to show the rate of photosynthesis by the plant leaf discs against the light intensity. **2**

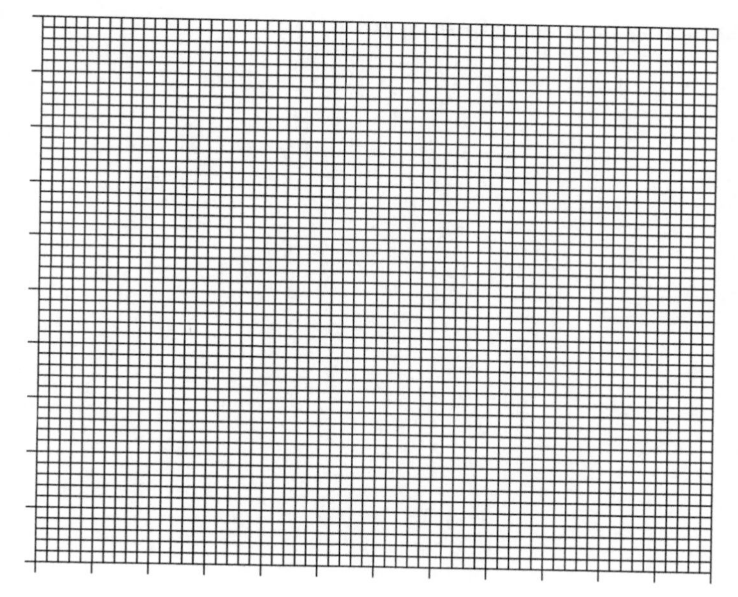

(b) From these results, draw a conclusion about the effect of light intensity on the rate of photosynthesis.

____

____ **2**

(c) Identify the independent variable in this investigation.

____ **1**

(d) Describe how the light intensity could be varied in this investigation.

____

____ **1**

19. Answer **either A or B**

A Describe social behaviour in animals under the following headings. **9**

(i) Altruism and kin selection

(ii) Primate behaviour

OR

B Give an account of genomic sequencing under the following headings.

(i) Description of genomic sequencing

(ii) Importance of genomic sequencing in personalised medicine

(iii) Importance of genomic sequencing in phylogenetics **9**

Labelled diagrams may be used where appropriate.

[END OF QUESTION PAPER]